HEINEMANN MODULAR MATHEMATICS
for
LONDON AS AND A-LEVEL
Pure Mathematics 1

Geoff Mannall   Michael Kenwood

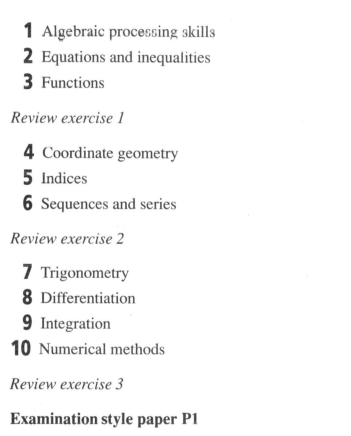

Heinemann Educational Publishers,
Halley Court, Jordan Hill, Oxford OX2 8EJ
a division of Reed Educational & Professional Publishing Ltd

MELBOURNE AUCKLAND FLORENCE PRAGUE MADRID ATHENS
SINGAPORE TOKYO SÃO PAULO CHICAGO PORTSMOUTH (NH)
MEXICO IBADAN GABORONE JOHANNESBURG
KAMPALA NAIROBI

First published 1994

97   12

ISBN 0 435 51807 0

Original design by Geoffrey Wadsley; additional design work by Jim Turner

Typeset and illustrated by TecSet Limited, Wallington, Surrey.

Printed in Great Britain by The Bath Press, Bath

Acknowledgements:

The publisher's and author's thanks are due to the University of London
Examinations and Assessment Council (ULEAC) for permission to reproduce
questions from past examination papers. These are marked with an [L].
   The answers have been provided by the authors and are not the responsibility
of the examining board.

About this book

This book is designed to provide you with the best preparation possible for your London Modular Mathematics P1 examination. The series authors are examiners and exam moderators themselves and have a good understanding of the exam board's requirements.

Finding your way around

To help to find your way around when you are studying and revising use the:

- **edge marks** (shown on the front page) – these help you to get to the right chapter quickly;
- **contents list** – this lists the headings that identify key syllabus ideas covered in the book so you can turn straight to them;
- **index** – if you need to find a topic the **bold** number shows where to find the main entry on a topic.

Remembering key ideas

We have provided clear explanations of the key ideas and techniques you need throughout the book. Key ideas you need to remember are listed in a **summary of key points** at the end of each chapter and marked like this in the chapters:

■
$$\sqrt{ab} = \sqrt{a} \cdot \sqrt{b}$$

Exercises and exam questions

In this book questions are carefully graded so they increase in difficulty and gradually bring you up to exam standard.

- **past exam questions** are marked with an L;
- **review exercises** on pages 52, 113 and 222 help you practise answering questions from several areas of mathematics at once, as in the real exam;
- **exam style practice paper** – this is designed to help you prepare for the exam itself;
- **answers** are included at the end of the book – use them to check your work.

Contents

5 Indices

6 Sequences and series

7 Trigonometry

8 Differentiation

9 Integration

10 Numerical methods

Algebraic processing skills

Algebra is the area of mathematics that uses symbols to represent numbers and to make generalisations about the relationships between them.

To succeed at an advanced level in mathematics you need to be able to use algebra confidently in a variety of situations. This chapter introduces some of the terminology of algebra and reviews some basic techniques of manipulating algebraic expressions – adding, subtracting, multiplying and factorising them.

1.1 Polynomials

A **polynomial** is an algebraic expression that is a sum of a number of terms. The name comes from the Greek words poly (many) and nomen (names or terms). Here is an example:

$$2x^3 + 7x^2 + 5x$$

These are each called **terms**

The most general form of a polynomial is written as:

$$a_n x^n + a_{n-1} x^{n-1} + \ldots + a_2 x^2 + a_1 x + a_0$$

This is called a polynomial in x of **degree** n, meaning the highest power of x is n. In a polynomial, n must be a positive integer (written $n \in \mathbb{Z}^+$) and a_n must not be zero ($a_n \neq 0$). So $2x^3 + 7x^2 + 5x + 2$ is a polynomial of degree 3.

Also $5x^6 + 2x^3 + 4x + 1$ is a polynomial of degree 6, since the expression could be written as

$$5x^6 + 0x^5 + 0x^4 + 2x^3 + 0x^2 + 4x + 1.$$

$2.3x^4 - 1.2x^2 + 7$ is a polynomial of degree 4, but

$$5x^3 + 6x - \frac{2}{x} - \frac{6}{x^5}$$

is *not* a polynomial, as it includes terms such as

$$\frac{2}{x} \text{ and } \frac{6}{x^5}$$

The number a_n is called the **coefficient** of x^n and, whilst n must be a positive integer, no restriction applies to a_n.

Adding and subtracting polynomials

To add or subtract two polynomials write one expression underneath the other and align corresponding terms – those in which x is raised to the same power. Leave gaps for missing terms.

Example 1
Add $3x^2 + 2x + 5$ to $6x^3 + x + 7$.

Write this as

$$
\begin{array}{r}
3x^2 + 2x + 5 \\
6x^3 \phantom{{}+3x^2} + x + 7 \\
\hline
6x^3 + 3x^2 + 3x + 12
\end{array}
$$

Example 2
Add $5x^7 - 2x^3 - 3$ to $6x^6 + 5x^5 - 3x^3 + 2$.

Write this as:

$$
\begin{array}{r}
5x^7 \phantom{{}+6x^6+5x^5} - 2x^3 - 3 \\
6x^6 + 5x^5 - 3x^3 + 2 \\
\hline
5x^7 + 6x^6 + 5x^5 - 5x^3 - 1
\end{array}
$$

Example 3
Subtract $2x^3 - 7x + 3$ from $4x^4 - 2x^2 + 3x - 2$.

Write this as:

$$
\begin{array}{r}
4x^4 \phantom{{}+2x^3} - 2x^2 + 3x - 2 \\
2x^3 \phantom{{}-2x^2} - 7x + 3 \\
\hline
4x^4 - 2x^3 - 2x^2 + 10x - 5
\end{array}
$$

When adding or subtracting polynomials it is usually easier to start with each polynomial in its own set of brackets, then remove the brackets and add or subtract the corresponding terms. If you use this method remember that a $+$ sign in front of a bracket leaves the signs inside the bracket unaltered when the bracket is removed, but a $-$ sign in front of the bracket changes each of the signs inside the bracket when the bracket is removed.

Example 4
Simplify $(2x^2 - 3x + 2) + (-3x^2 + 2x - 6)$.

$$
\begin{aligned}
&(2x^2 - 3x + 2) + (-3x^2 + 2x - 6) \\
&= 2x^2 - 3x + 2 - 3x^2 + 2x - 6 \\
&= -x^2 - x - 4
\end{aligned}
$$

Example 5
Add $(3x^2 - 2)$, $(6x^3 - 2x^2 + 1)$, $(4x^2 - 2x + 3)$ and $(6x - 7)$.

$$
\begin{aligned}
&(3x^2 - 2) + (6x^3 - 2x^2 + 1) + (4x^2 - 2x + 3) + (6x - 7) \\
&= 3x^2 - 2 + 6x^3 - 2x^2 + 1 + 4x^2 - 2x + 3 + 6x - 7 \\
&= 6x^3 + 5x^2 + 4x - 5
\end{aligned}
$$

Example 6
Subtract $(8x^4 - 2x^2 - 3x)$ from $(6x^3 - 7x + 2)$.

$$
\begin{aligned}
&(6x^3 - 7x + 2) - (8x^4 - 2x^2 - 3x) \\
&= 6x^3 - 7x + 2 - 8x^4 + 2x^2 + 3x \\
&= -8x^4 + 6x^3 + 2x^2 - 4x + 2
\end{aligned}
$$

Example 7
Simplify $(2x^3 + 6x - 2) - (3x^2 - 4x + 7) - (6x - 3) + (2x^2 + 5x)$.

Removing the brackets:

$$
\begin{aligned}
&2x^3 + 6x - 2 - 3x^2 + 4x - 7 - 6x + 3 + 2x^2 + 5x \\
&= 2x^3 - x^2 + 9x - 6
\end{aligned}
$$

Multiplying one polynomial by another
You can use similar methods to multiply two polynomials.

Example 8

Multiply $x^3 - 2x + 4$ by $3x - 7$.

Write it as:

$$
\begin{array}{r}
x^3 \qquad\quad -2x + 4 \\
3x - 7
\end{array}
$$

Multiplying by $3x$: $\quad 3x^4 \qquad\quad -6x^2 + 12x$

Multiplying by -7: $\qquad\quad -7x^3 \qquad\quad +14x - 28$

Adding: $\qquad\qquad 3x^4 - 7x^3 - 6x^2 + 26x - 28$

Example 9

Multiply $2x^5 + 7x^3 + 2$ by $8x^3 - 2x - 3$.

Write:

$$
\begin{array}{r}
2x^5 \qquad\quad +7x^3 \qquad +2 \\
8x^3 - 2x - 3
\end{array}
$$

Multiplying by $8x^3$: $\;\; 16x^8 + 56x^6 \qquad\qquad +16x^3$

Multiplying by $-2x$: $\qquad\quad -4x^6 \qquad -14x^4 \qquad -4x$

Multiplying by -3: $\qquad\qquad\quad -6x^5 \qquad -21x^3 \quad -6$

Adding: $\qquad\quad 16x^8 + 52x^6 - 6x^5 - 14x^4 - 5x^3 - 4x - 6$

As with addition and subtraction, when multiplying polynomials it is usually quicker to start with each polynomial in its own set of brackets, then remove the brackets and collect corresponding terms. Once again, remember that a + sign outside the bracket leaves the signs inside the bracket unaltered and a − sign outside the bracket changes all the signs inside the bracket.

Example 10

Multiply $(2x^2 - 3x + 4)$ by $(3x^3 - x + 1)$.

$$
\begin{aligned}
&(2x^2 - 3x + 4)(3x^3 - x + 1) \\
={}&2x^2(3x^3 - x + 1) - 3x(3x^3 - x + 1) + 4(3x^3 - x + 1) \\
={}&6x^5 - 2x^3 + 2x^2 - 9x^4 + 3x^2 - 3x + 12x^3 - 4x + 4 \\
={}&6x^5 - 9x^4 + 10x^3 + 5x^2 - 7x + 4
\end{aligned}
$$

Example 11

Multiply $(5x^2 - 2x + 3)$ by $(2x^3 - 4x^2 + 2x - 3)$.

$$
\begin{aligned}
&(5x^2 - 2x + 3)(2x^3 - 4x^2 + 2x - 3) \\
={}&5x^2(2x^3 - 4x^2 + 2x - 3) - 2x(2x^3 - 4x^2 + 2x - 3) + 3(2x^3 - 4x^2 + 2x - 3) \\
={}&10x^5 - 20x^4 + 10x^3 - 15x^2 - 4x^4 + 8x^3 - 4x^2 + 6x + 6x^3 - 12x^2 + 6x - 9 \\
={}&10x^5 - 24x^4 + 24x^3 - 31x^2 + 12x - 9
\end{aligned}
$$

Exercise 1A

1 Add $(5x^7 + 2x^5 + 3x^2)$ and $(6x^7 + 2x^6 + 3x^5 + 2x)$
2 Add $(3x^5 - 2x^2 + 6x)$ and $(4x^4 - 2x^3 + 3x^2 + 2)$
3 Add $(2x^3 - 3x + 1)$, $(2x^2 + x - 6)$ and $(2x - 3)$
4 Add $(8x^4 - 3x^2 + 6x + 1)$, $(5x^3 - 4x^2 - 2x + 2)$
   and $(-2x^4 + 6x - 3)$
5 Simplify $(5x^6 - 2x^3 + 7x) + (3x^5 - 2x^4 + 3x^3 + 7)$
6 Simplify $(-3x^4 + 2x^2 + 7x + 2) + (8x^3 - 3x^2 + 2x - 5)$
7 Simplify $(5x^7 - 6x^3 + 7) + (-6x^6 + 2x^3 - 9) + (5x^4 - 2x - 3)$
8 Simplify
   $(2x^2 - 3) + (3x^2 + 2x - 1) + (-2x^3 - 2x^2 + 2x) + (6x - 3)$
9 Subtract $(4x^5 - 2x^3 + 7)$ from $(5x^5 + 7x - 2)$
10 Subtract $(3x^3 + 7x^2 + 2)$ from $(6x^3 + 7x + 3)$
11 Subtract $(2x^5 - 3x^3 - x - 2)$ from $(6x^3 - 9x^2 + 2x - 3)$
12 Subtract $(-2x^3 - 4x^2 - 3x + 2)$ from
    $(-3x^5 + 2x^3 - 4x^2 + 6x - 7)$
13 Simplify $(6x^4 + 3x^2 - 2x + 9) - (3x^3 - 7x^2 - 6x + 7)$
14 Simplify $(5x^3 - 2x + 1) - (2x^3 - 7x + 3)$
15 Simplify $(6x^5 - 2x + 2) - (3x^3 - 2x - 5) - (7x + 4)$
16 Simplify $(4x^6 - 2x + 1) + (3x^3 + 7x - 4) - (-3x^6 - 2x + 7)$
17 Simplify
    $(8x^4 - 2x^2 + 7) - (6x^3 - 3x^2 - 1) + (6x + 4) - (3x^2 - 2x - 3)$
18 Multiply $(3x^2 + 2x + 4)$ by $(4x^3 + 7x + 6)$
19 Multiply $(2x^3 - 6x - 3)$ by $(5x^4 + 2x^2 - x)$
20 Multiply $(4x^2 + 2x - 1)$ by $(6x^2 - x + 7)$
21 Multiply $(3x^3 - 2x - 6)$ by $(2x^2 - 3x + 1)$
22 Simplify $(8x^5 - 2x^4 + 3x - 7)(2x^3 - 3x^2 - 9x + 6)$
23 Simplify $(2x^4 - 7x + 6)(x^3 - 2x^2 + 7x - 3)$
24 Simplify $(3x^2 - 2x - 4)(8x^5 - 3x^3 - 7x + 3)$
25 Simplify $(-7x + 2x^2 - 6x^3)(-x^4 + 6x^2 - 2x - 5)$

1.2 Factorising polynomials

From the work done so far you should be able to show that:

$$(2x^2 - 1)(3x + 4) \text{ is equivalent to } 6x^3 + 8x^2 - 3x - 4.$$

Because these two expressions are equal for *all* values of x (not just for some values of x as in the case of equations) we say that:

$(2x^2 - 1)(3x + 4)$ **is identically equal to** $6x^3 + 8x^2 - 3x - 4$

$(2x^2 - 1)(3x + 4) \equiv 6x^3 + 8x^2 - 3x - 4$ is an **identity**. The identity symbol is $\equiv$.

You should by now be able to check that the following are identities (that is, they are true for all values of x and y).

$(x + y)^2 \equiv (x + y)(x + y) \equiv x^2 + 2xy + y^2$

$(x - y)^2 \equiv (x - y)(x - y) \equiv x^2 - 2xy + y^2$

$(x - y)(x + y) \equiv x^2 - y^2$

$(x + y)^3 \equiv (x + y)(x + y)(x + y) \equiv x^3 + 3x^2y + 3xy^2 + y^3$

$(x - y)^3 \equiv (x - y)(x - y)(x - y) \equiv x^3 - 3x^2y + 3xy^2 - y^3$

The **factors** of each expression are shown on the left-hand side. Multiplying the factors together gives the full expression. So $(x - y)$ and $(x + y)$ are the factors of $x^2 - y^2$. The process of taking an expression and writing it as a product of its factors is called **factorisation**. So if you factorise $x^2 - y^2$ you get $(x - y)(x + y)$.

The simplest type of factor is the **common factor**. As its name suggests, it is a factor which is common to each term in a polynomial.

Example 12
Factorise $4x^3 + 7x$.

Each term in the polynomial contains the factor x so
$4x^3 + 7x \equiv x(4x^2 + 7)$.

Example 13
Factorise $3x^3 + 6x^2 - 12x$.

Each term in the polynomial contains the factor $3x$ so
$3x^3 + 6x^2 - 12x \equiv 3x(x^2 + 2x - 4)$.

Many other polynomials can be factorised using the five identities given above.

Example 14
Factorise $4x^2 + 12x + 9$.

If you write $4x^2 + 12x + 9$ as $(2x)^2 + 2(2x)(3) + 3^2$, then you can compare it with the identity $X^2 + 2XY + Y^2 \equiv (X + Y)^2$. In this case $X = 2x$ and $Y = 3$.

So:
$$4x^2 + 12x + 9 \equiv (2x + 3)^2$$

Example 15

Factorise $4x^2 - 1$.

If $4x^2 - 1$ is written as $(2x)^2 - 1^2$ then it may be compared with the identity $X^2 - Y^2 \equiv (X - Y)(X + Y)$. Here, $X = 2x$ and $Y = 1$. So

$$4x^2 - 1 \equiv (2x - 1)(2x + 1)$$

Example 16

Factorise $x^3 - 12x^2 + 48x - 64$.

Write $x^3 - 12x^2 + 48x - 64$ as $x^3 - 3x^2(4) + 3x(4)^2 - (4)^3$.

You can then compare it with the standard identity

$$(X - Y)^3 \equiv X^3 - 3X^2Y + 3XY^2 - Y^3$$

where $X = x$ and $Y = 4$.

So:
$$x^3 - 12x^2 + 48x - 64 \equiv (x - 4)^3$$

Exercise 1B

Factorise:

1 $2x^2 - 3x^3$

2 $5y^5 + 2y^2 + 7y$

3 $2x^2 + 18x$

4 $40y - 8$

5 $6xy + 18x$

6 $5xy^3 + 15x^2y$

7 $2x - 6y + 10xy$

8 $x^3 + 3x^2 - 6x$

9 $6pq^2 + 9p^2q$

10 $2c(c + d) + 4cd$

11 $x^2 - 36$

12 $y^2 - 81$

13 $100 - b^2$

14 $8a^2 - 18b^2$

15 $18 - 50b^2$

16 $a^2 - a^2b^2$

17 $2x^2 - 18$

18 $3x^2 - 27$

19 $17a^2 - 68$

20 $48 - 147c^2$

21 $x^2 + 6x + 9$

22 $x^2 - 8x + 16$

23 $x^2 + 14x + 49$

24 $x^2 - 18x + 81$

25 $3x^2 - 36x + 108$

26 $9x^2 + 30x + 25$

27 $4x^2 - 28x + 49$

28 $45x^2 - 30x + 5$

29 $6x^2 + 12x + 2$

30 $25x^2 - 30x + 9$

31 $25x^2 + 20x + 4$
32 $9x^2 - 42x + 49$
33 $x^3 + 6x^2 + 12x + 8$
34 $x^3 - 9x^2 + 27x - 27$
35 $8x^3 + 12x^2 + 6x + 1$
36 $27x^3 - 27x^2 + 9x - 1$
37 $8x^3 + 36x^2 + 54x + 27$
38 $8x^3 - 60x^2 + 150x - 125$
39 $x^3y + 6x^2y + 12xy + 8y$
40 $3x^3 - 27x^2 + 81x - 81$

1.3 Factorising trinomials

A polynomial of the form $ax^2 + bx + c$ is called a **trinomial** as it contains three terms. Many trinomials can be factorised. But not many can be factorised as a perfect square using

$$(x + y)^2 \equiv x^2 + 2xy + y^2 \text{ or } (x - y)^2 \equiv x^2 - 2xy + y^2$$

These are special cases.

However, you should be able to show that:

$$(x + a)(x + b) \equiv x^2 + (a + b)x + ab$$

Polynomials of degree two which contain three terms (i.e. trinomials) and which have the coefficient of x^2 as 1 can often be factorised using this identity.

Example 17
Factorise $x^2 + 5x + 6$.

If you compare $x^2 + 5x + 6$ with $x^2 + (a + b)x + ab$, you can see that $ab = 6$. If you assume that a and b are integers then either $a = 6$ and $b = 1$, or $a = 1$ and $b = 6$, or $a = 2$ and $b = 3$, or $a = 3$ and $b = 2$. If you take $a = 2$ and $b = 3$ then the middle term, $(a + b)x$, becomes $5x$, as required.

So: $\qquad\qquad x^2 + 5x + 6 \equiv (x + 2)(x + 3)$

Example 18
Factorise $x^2 + 3x - 28$.

Compare $x^2 + 3x - 28$ with $x^2 + (a + b)x + ab$, and you see that $ab = -28$. So $a = -28, b = 1$ or $a = 28, b = -1$, or $a = -2, b = 14$ or $a = 2, b = -14$ or $a = -4, b = 7$ or $a = 4, b = -7$. If you choose $a = -4, b = 7$ you find that $(a + b)x$ equates to $3x$ as required.

So: $\qquad\qquad x^2 + 3x - 28 \equiv (x - 4)(x + 7)$

Exercise 1C

Factorise:

1 $x^2 + 9x + 14$	**2** $x^2 - 8x + 15$	**3** $x^2 - x - 12$
4 $x^2 + 13x + 12$	**5** $x^2 - 9x - 10$	**6** $x^2 - 3x + 2$
7 $x^2 + 15x + 56$	**8** $x^2 + 4x - 12$	**9** $x^2 + 7x + 10$
10 $x^2 - 3x - 40$	**11** $x^2 - 7x + 12$	**12** $x^2 - 14x + 45$
13 $x^2 + 11x + 18$	**14** $x^2 - 10x + 21$	**15** $x^2 + 9x + 20$
16 $x^2 - 17x + 30$	**17** $x^2 - 22x + 120$	**18** $x^2 + 3x - 180$
19 $x^2 - x - 110$	**20** $x^2 + 11x - 620$	

Trinomials that need to be factorised often have a coefficient of x^2 that is *not* 1. In such cases you must use the identity

$$(ax + b)(cx + d) \equiv acx^2 + (ad + bc)x + bd$$

Example 19

Factorise $6x^2 - 8x - 8$.

Comparing $6x^2 - 8x - 8$ with $acx^2 + (ad + bc)x + bd$, you can see that $ac = 6$ and $bd = -8$.

By trying different possible numbers for a, b, c and d you find that if $a = 2$ and $c = 3$ (to give $ac = 6$), and if $b = -4$ and $d = 2$ (to give $bd = -8$) then the middle term, $(ad + bc)x$, becomes

$$[(2 \times 2) + (-4 \times 3)]x \text{ or } -8x$$

as required.

So: $$6x^2 - 8x - 8 \equiv (2x - 4)(3x + 2)$$

Example 20

Factorise $15x^2 - 11x - 12$.

Compare $15x^2 - 11x - 12$ with $acx^2 + (ad + bc)x + bd$.

Then $ac = 15$ and $bd = -12$.

If $a = 3$ and $c = 5$ (to give $ac = 15$), and if $b = -4$ and $d = 3$ (to give $bd = -12$) then $(ad + bc)x$ becomes

$$[(3 \times 3) + (-4 \times 5)]x = -11x$$

as required.

So: $$15x^2 - 11x - 12 \equiv (3x - 4)(5x + 3)$$

Sometimes a combination of the above methods is required in order to factorise a polynomial. This, at first sight, might seem difficult.

However, in this module you will only be asked to factorise polynomials of degree at most 3, so things are not as bad as they might at first appear, as the next example shows.

Example 21

Factorise $x^3 + 4x^2 + 3x$.

The trinomial $x^3 + 4x^2 + 3x$ has a common factor of x.

So:
$$x^3 + 4x^2 + 3x \equiv x(x^2 + 4x + 3)$$

Now
$$x^2 + 4x + 3 \equiv (x + 3)(x + 1)$$

So:
$$x^3 + 4x^2 + 3x \equiv x(x + 3)(x + 1)$$

Exercise 1D

Factorise:

1 $2x^2 + 7x + 6$	**2** $2x^2 + 3x - 9$	**3** $3x^2 + 7x + 2$
4 $5x^2 - 14x - 3$	**5** $3y^2 + 13y + 14$	**6** $2y^2 - 9y - 5$
7 $7y^2 - 30y + 8$	**8** $5y^2 - 27y + 10$	**9** $3y^2 - 10y - 25$
10 $3y^2 + 4y - 4$	**11** $4x^2 + 8x + 3$	**12** $6x^2 - x - 2$
13 $8x^2 + 14x - 15$	**14** $10x^2 - 23x + 12$	**15** $12x^2 - 5x - 3$
16 $10x^2 + 11x - 35$	**17** $12x^2 - 11x + 2$	**18** $9x^2 + 21x + 10$
19 $4x^2 + 8x - 21$	**20** $35x^2 + x - 6$	**21** $x^3 - 4x^2 - 21x$
22 $6x - 7x^2 + 2x^3$	**23** $10x^5 + 9x^4 + 2x^3$	**24** $12 - 4x - 40x^2$
25 $6x^2y + 19xy - 7y$		

1.4 Surds

Some numbers which contain a root sign can be evaluated exactly. For example, $\sqrt{4} = 2$, $\sqrt{100} = 10$, $\sqrt[3]{27} = 3$, $\sqrt[5]{\frac{1}{32}} = \frac{1}{2}$ – they can be written as exact numbers.

This is because each of these numbers is **rational**. However, some numbers which contain a root sign cannot be evaluated exactly because they are **irrational** – they cannot be written exactly as a decimal. Hence $\sqrt{3} = 1.732\,05\ldots$ and $\sqrt[3]{7} = 1.9129\ldots$ These decimals go on for ever. Such numbers are called **surds**.

The answers to many mathematical problems appear more elegant than would otherwise be the case if they are given as surds. Also, when an answer to a problem is given as a surd, it is the *exact* answer to the problem. If the answer to a problem is irrational and is given in decimal form then it can only be given to a specified number of decimal places and so it can only be an approximation to the correct answer. For both these reasons many mathematical questions require the answer to be given as a surd. So you need to learn how to manipulate surds.

Manipulating surds

Here are the rules for the manipulation of surds:

- $\sqrt{ab} = \sqrt{a} \cdot \sqrt{b}$

- $\sqrt{\dfrac{a}{b}} = \dfrac{\sqrt{a}}{\sqrt{b}}$

- $a\sqrt{b} + c\sqrt{b} = (a+c)\sqrt{b}$

- $a\sqrt{b} - c\sqrt{b} = (a-c)\sqrt{b}$

Notice that $\sqrt{a} + \sqrt{b}$ is *not* the same as $\sqrt{(a+b)}$.

For example, $\sqrt{25} + \sqrt{144} = 5 + 12 = 17$, which is not the same as

$$\sqrt{(25 + 144)} = \sqrt{169} = 13$$

Example 22
Simplify $\sqrt{80}$.

$$\sqrt{80} = \sqrt{(16 \times 5)} = \sqrt{16} \times \sqrt{5} = 4\sqrt{5}$$

Example 23

Simplify $\dfrac{\sqrt{63}}{3}$.

$$\frac{\sqrt{63}}{3} = \frac{\sqrt{(9 \times 7)}}{3} = \frac{\sqrt{9} \times \sqrt{7}}{3} = \frac{3 \times \sqrt{7}}{3} = \sqrt{7}$$

Example 24
Simplify $\sqrt{75} + 2\sqrt{48} - 5\sqrt{12}$.

$$\begin{aligned}
&\sqrt{75} + 2\sqrt{48} - 5\sqrt{12} \\
&= \sqrt{(25 \times 3)} + 2\sqrt{(16 \times 3)} - 5\sqrt{(4 \times 3)} \\
&= (\sqrt{25} \times \sqrt{3}) + 2(\sqrt{16} \times \sqrt{3}) - 5(\sqrt{4} \times \sqrt{3}) \\
&= 5\sqrt{3} + 8\sqrt{3} - 10\sqrt{3} \\
&= 3\sqrt{3}
\end{aligned}$$

Rationalising the denominator

When the denominator of a fraction is a surd, it is normal to try to remove the surd from the denominator. This process is called 'rationalising the denominator'. Before you can do this you need to learn three more rules:

- If a fraction is in the form $\dfrac{1}{\sqrt{a}}$, then multiply both the top and bottom of the fraction by $\sqrt{a}$.

- If a fraction is the form $\dfrac{1}{a + \sqrt{b}}$, then multiply both the top and the bottom of the fraction by $(a - \sqrt{b})$.

- If a fraction is in the form $\dfrac{1}{a - \sqrt{b}}$, then multiply both the top and bottom of the fraction by $(a + \sqrt{b})$.

The following examples demonstrate how to rationalise the denominator by applying these three rules.

Example 25
Rationalise the denominator of $\dfrac{1}{\sqrt{3}}$.

$$\frac{1}{\sqrt{3}} = \frac{1 \times \sqrt{3}}{\sqrt{3} \times \sqrt{3}} = \frac{\sqrt{3}}{3}$$

Example 26
Rationalise the denominator of $\dfrac{1}{2 - \sqrt{3}}$

$$\frac{1}{2 - \sqrt{3}} = \frac{1 \times (2 + \sqrt{3})}{(2 - \sqrt{3})(2 + \sqrt{3})} = \frac{2 + \sqrt{3}}{4 - 3}$$
$$= 2 + \sqrt{3}$$

Example 27
Rationalise the denominator of $\dfrac{\sqrt{3} - \sqrt{2}}{\sqrt{3} + \sqrt{2}}$.

$$\frac{\sqrt{3} - \sqrt{2}}{\sqrt{3} + \sqrt{2}} = \frac{(\sqrt{3} - \sqrt{2})(\sqrt{3} - \sqrt{2})}{(\sqrt{3} + \sqrt{2})(\sqrt{3} - \sqrt{2})}$$
$$= \frac{3 - 2\sqrt{2}\sqrt{3} + 2}{3 - 2}$$
$$= 5 - 2\sqrt{6}$$

Exercise 1E

Simplify:

1 $\sqrt{27}$ **2** $\sqrt{45}$ **3** $\sqrt{162}$

4 $\sqrt{48}$ **5** $\sqrt{75}$ **6** $\sqrt{147}$

7 $\sqrt{567}$ **8** $\sqrt{112}$ **9** $\dfrac{\sqrt{12}}{2}$

10 $\dfrac{\sqrt{98}}{7}$ **11** $\dfrac{\sqrt{18}}{\sqrt{2}}$ **12** $\dfrac{\sqrt{27}}{\sqrt{3}}$

13 $\sqrt{12} + 3\sqrt{75}$ **14** $\sqrt{200} + \sqrt{18} - 2\sqrt{72}$

15 $\sqrt{20} + 2\sqrt{45} - 3\sqrt{80}$ **16** $5\sqrt{6} - \sqrt{24} + \sqrt{294}$

17 $\sqrt{63} - 2\sqrt{28} + \sqrt{175}$

Rationalise the denominators:

18 $\dfrac{1}{\sqrt{2}}$ **19** $\dfrac{1}{\sqrt{7}}$ **20** $\dfrac{7}{\sqrt{5}}$

21 $\dfrac{\sqrt{2}}{3\sqrt{3}}$ **22** $\dfrac{\sqrt{8}}{\sqrt{32}}$ **23** $\dfrac{\sqrt{5}}{\sqrt{45}}$

24 $\dfrac{\sqrt{3}}{\sqrt{21}}$ **25** $\dfrac{\sqrt{11}}{\sqrt{132}}$ **26** $\dfrac{1}{1 - \sqrt{2}}$

27 $\dfrac{1}{\sqrt{3} - 1}$ **28** $\dfrac{2}{\sqrt{5} - 1}$ **29** $\dfrac{3}{4 - \sqrt{7}}$

30 $\dfrac{1}{3 - \sqrt{7}}$ **31** $\dfrac{1}{\sqrt{11} - 4}$ **32** $\dfrac{1}{\sqrt{7} - \sqrt{3}}$

33 $\dfrac{2}{\sqrt{11} - \sqrt{5}}$ **34** $\dfrac{7}{\sqrt{13} - \sqrt{3}}$ **35** $\dfrac{7}{2 + \sqrt{7}}$

36 $\dfrac{2 - \sqrt{3}}{\sqrt{11} - 4}$ **37** $\dfrac{7\sqrt{2}}{\sqrt{8} - \sqrt{7}}$ **38** $\dfrac{\sqrt{5} - \sqrt{3}}{\sqrt{5} + \sqrt{3}}$

39 $\dfrac{\sqrt{13} - \sqrt{7}}{\sqrt{13} + \sqrt{7}}$ **40** $\dfrac{\sqrt{43} - \sqrt{23}}{\sqrt{43} + \sqrt{23}}$

SUMMARY OF KEY POINTS

1 A polynomial in x of degree n is an expression such as
$a_n x^n + a_{n-1} x^{n-1} + \ldots + a_2 x^2 + a_1 x + a_0$,
where $n \in \mathbb{Z}^+$, $a_n \neq 0$.

2 $(x + y)^2 \equiv x^2 + 2xy + y^2$

3 $(x - y)^2 \equiv x^2 - 2xy + y^2$

4 $(x - y)(x + y) \equiv x^2 - y^2$

5 $(x + y)^3 \equiv x^3 + 3x^2y + 3xy^2 + y^3$

6 $(x - y)^3 \equiv x^3 - 3x^2y + 3xy^2 - y^3$

7 A surd is an irrational number containing a root.

8 $\sqrt{ab} = \sqrt{a} \cdot \sqrt{b}$

9 $\sqrt{\dfrac{a}{b}} = \dfrac{\sqrt{a}}{\sqrt{b}}$

10 $a\sqrt{b} \pm c\sqrt{b} = (a \pm c)\sqrt{b}$

11 To rationalise a denominator when it is in the form $\dfrac{1}{\sqrt{a}}$, multiply both top and bottom of the fraction by $\sqrt{a}$.

12 To rationalise a denominator when it is in the form $\dfrac{1}{a \pm \sqrt{b}}$, multiply both top and bottom of the fraction by $(a \mp \sqrt{b})$.

Equations and inequalities

Equations occur throughout advanced courses in mathematics. This chapter revises some of the techniques for solving equations that you may have met at GCSE, and shows you how to solve some other types of equations and inequalities.

2.1 Linear equations in one unknown

An equation in x which can be written in the form $ax + b = 0$, with the highest power of x being 1, is called a **linear equation** or an equation of the first degree. (You might like to compare this with the degree of a polynomial described on p. 1.)

Linear equations have just one unknown and the value of this unknown is the solution or **root** of the equation. The method of solution is to get all the terms in x onto the left-hand side (LHS) of the equation and everything else onto the right-hand side (RHS) of the equation.

Example 1

Solve $5x + 6 = 2x + 12$.

$5x + 6 = 2x + 12$ can be written

$$5x - 2x = 12 - 6$$
$$3x = 6$$
$$x = 2$$

You can check this solution by putting $x = 2$ in the original equation:

LHS $= (5 \times 2) + 6 = 10 + 6 = 16$
RHS $= (2 \times 2) + 12 = 4 + 12 = 16$

Example 2

Solve $4x - 2 + 6x + 3 = 5x - 22 + 2x - 1$.

$4x - 2 + 6x + 3 = 5x - 22 + 2x - 1$ can be written

$$4x + 6x - 5x - 2x = -22 - 1 + 2 - 3$$
$$3x = -24$$
$$x = -8$$

Check:

$$\begin{aligned} \text{LHS} &= (4 \times -8) - 2 + (6 \times -8) + 3 \\ &= -32 - 2 - 48 + 3 = -79 \\ \text{RHS} &= (5 \times -8) - 22 + (2 \times -8) - 1 \\ &= -40 - 22 - 16 - 1 = -79 \end{aligned}$$

Example 3

Solve $\dfrac{x - 3}{4} - \dfrac{2 - x}{3} = 5$.

Multiply by 12, which is the common denominator of the fractions.

$$3(x - 3) - 4(2 - x) = 5 \times 12$$
$$3x - 9 - 8 + 4x = 60$$
$$3x + 4x = 60 + 9 + 8$$
$$7x = 77$$
$$x = 11$$

Check:

$$\text{LHS} = \frac{11 - 3}{4} - \frac{2 - 11}{3} = \frac{8}{4} - \frac{-9}{3} = 2 + 3 = 5 = \text{RHS}$$

2.2 Simultaneous linear equations in two unknowns

An equation in two unknowns x and y that can be written in the form $ax + by + c = 0$ is also linear since the highest powers of x and y are each 1.

Unlike a linear equation in one unknown which has only one solution, a linear equation in two unknowns has many solutions.

For example, here are some of the solutions of the equation $3x + 2y = 6$:

$$x = 0 \quad \text{and} \quad y = 3, \quad x - 1 \quad \text{and} \quad y = 1\tfrac{1}{2}, \quad x = 2 \quad \text{and} \quad y = 0$$

In fact a linear equation in two unknowns x and y has an infinite number of solutions. But if you have another linear equation in x and y there will usually be only one value of x and one value of y that work for both equations. These values are called the solution. The solution will satisfy both equations simultaneously (at the same time) so the equations are called **simultaneous linear equations.**

You need to be sure that the second equation is not of the form:

$$max + mby + d = 0$$

otherwise there will not be a unique solution.

There are three common methods of solving simultaneous linear equations – by graphing, by substitution, and by elimination.

Solving simultaneous linear equations graphically

Any equation $ax + by + c = 0$ is linear and has a straight line (linear) graph.

Example 4
Solve these simultaneous linear equations:

$$x + y = 4$$
$$4x - y = 1$$

To solve these equations graphically, first draw up a table of values for each equation. Since you know that each graph will be a straight line, the minimum number of points that you need to plot each graph is two. However, four is the safe minimum; if you make a mistake in calculating one or more of the points then they will not lie on a straight line and so your mistake will be obvious.

$$x + y = 4$$

x	0	2	3	4
y	4	2	1	0

$$4x - y = 1$$

x	0	2	3	4
y	-1	7	11	15

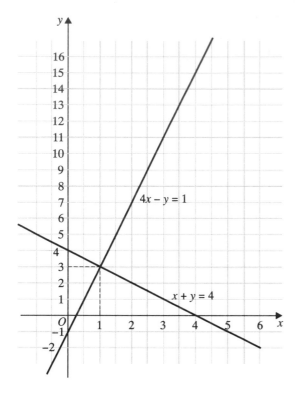

For the solution, read off the x- and y-coordinates at the point of intersection of the two lines. Here they meet at $x = 1$ and $y = 3$. You should always check the solution to a pair of simultaneous equations by putting the solution back into the equations to make sure that it satisfies both of them. In this case $x + y = 1 + 3 = 4$, as required and $4x - y = 4 - 3 = 1$, as required. So $x = 1$, $y = 3$ is indeed the solution of the pair of simultaneous equations.

Solving simultaneous linear equations by substitution

The second method of solving a pair of simultaneous linear equations is by substitution. To use this method, rearrange one of the equations to make either x or y the subject of the equation and then substitute the resulting expression into the other equation.

Example 5
Solve these simultaneous equations by the method of substitution:

$$x + y = 4$$
$$4x - y = 1$$

Taking the equation $x + y = 4$, make y the subject by taking x to the other side of the equation. So $y = 4 - x$. Then substitute this expression for y into the other equation.

So the second equation, $4x - y = 1$, becomes

$$4x - (4 - x) = 1$$
$$4x - 4 + x = 1$$
$$5x = 1 + 4$$
$$x = 1$$

If you substitute this value of x into the first equation you can evaluate y. So $y = 4 - 1 = 3$ and once again the solution is $x = 1$, $y = 3$, which can be checked mentally in the second equation.

Example 6
Solve these simultaneous linear equations by the method of substitution:

$$2x + 3y = 7$$
$$5x + 2y = 23$$

Making x the subject of the first equation gives $2x = 7 - 3y$ and so

$$x = \frac{7}{2} - \frac{3y}{2}$$

Substitute this into the second equation:

$$5\left(\frac{7}{2} - \frac{3y}{2}\right) + 2y = 23$$

so

$$\frac{35}{2} - \frac{15y}{2} + 2y = 23$$

or

$$17\tfrac{1}{2} - 7\tfrac{1}{2}y + 2y = 23$$
$$-5\tfrac{1}{2}y = 5\tfrac{1}{2}$$
$$y = -1$$

Substituting this value of y into the first equation gives

$$2x - 3 = 7$$

This is

$$2x = 10$$
$$x = 5$$

So the solution is $x = 5$ and $y = -1$.

As a check, put these values into the second equation (since this is not the one into which you have just substituted in order to find the value of x).

$$5x + 2y = (5 \times 5) + (2 \times -1) = 25 - 2 = 23$$

as required.

Solving simultaneous linear equations by elimination

Elimination is the most popular method of solving two linear simultaneous equations. It involves first equalising the number of xs or ys in each question and then eliminating them by either adding or subtracting the two equations as appropriate.

Example 7

Solve these simultaneous linear equations by the method of elimination:

$$2x + 3y = 7$$
$$5x + 2y = 23$$

To equalise the ys, multiply the first equation by 2 to get $4x + 6y = 14$ and then multiply the second equation by 3 to get $15x + 6y = 69$.

You can decide whether to add or subtract the two equations so as to eliminate the ys by applying a simple rule: if the signs are the same then subtract, but if the signs are different, then add. In this case the equal terms $6y$, which you are trying to eliminate, have the same sign $(+)$, so you subtract the two equations:

$$4x + 6y = 14$$
$$15x + 6y = 69$$

Subtracting:
$$-11x = -55$$
$$x = 5$$

Substituting this into the first equation gives:

$$10 + 3y = 7$$

i.e.
$$3y = -3$$
$$y = -1$$

and substituting into the second equation checks out as before. So, once again, you obtain the solution $x = 5$ and $y = -1$.

Example 8

Solve these simultaneous linear equations by the method of elimination:

$$2x + 3y = 13$$
$$7x - 5y = -1$$

Try equalising the ys in the two equations.

Multiply the first equation by 5:

$$10x + 15y = 65$$

Multiply the second equation by 3:

$$21x - 15y = -3$$

Since the y terms have different signs, add the two equations to obtain

$$31x = 62$$
$$x = 2$$

Substitute into the first equation:

$$4 + 3y = 13$$
$$3y = 9$$
$$y = 3$$

The solution is $x = 2$ and $y = 3$.

Check by substituting these values into the second equation. $(7 \times 2) - (5 \times 3) = 14 - 15 = -1$, as required.

Exercise 2A

Solve these equations:

1 $5x - 3 = 2x + 15$

2 $8x - 6 + 7x = 5x - 4 - 20$

3 $3(x + 4) + 2 = 2(4x + 3) - 2$

4 $5(x + 3) + 4(2x - 3) = 2(2x + 15)$

5 $2 + 7x - 4 - 2(3 + x) = 7 - 10x$

6 $3(2 + 3x) - 4(5 - 3x) = 28$

7 $\dfrac{x}{2} - \dfrac{x}{3} = 5$

8 $\dfrac{x}{2} - \dfrac{1}{3} = \dfrac{1}{4} - \dfrac{x}{5}$

9 $\dfrac{2}{x} - \dfrac{5}{2x} = 2$

10 $\dfrac{x + 2}{3} + \dfrac{2x + 1}{5} = 6$

11 $\dfrac{x+2}{3} - \dfrac{2x+1}{5} = 2$ **12** $\dfrac{3x-2}{4} - \dfrac{2x-3}{3} = \dfrac{x-1}{5}$

Solve these simultaneous linear equations:

13 $2x + y = 18$ **14** $x - 2y = 5$
         $x - 2y = -1$ $3x + y = 8$

15 $x + 2y = 4$ **16** $5x + 2y = -30$
         $3x + 5y = 9$ $3x + 4y = -32$

17 $14x + 4y = -1$ **18** $3x - 2y = 7$
         $-3x + 5y = 9$ $2x - 5y = 12$

19 $7p + 3q = 5$ **20** $6x - 11y = 4$
         $5p - q = 2$ $5x + 3y = -21$

21 $17x + 9y = 20$ **22** $y = 7z$
         $5x - 2y = -22$ $8y + 100 = 6z$

23 $2p + 3q = -5$ **24** $5a - 6b = 7$
         $3p - 5q = 21$ $8b - a = 2$

25 $2x - 7y = 57$ **26** $3x - 2y = 2$
         $3y - 11x = 6$ $9x - 4y = 1$

27 $24x + 12y + 7 = 0$
         $6x + 12y = 5$

2.3 Quadratic equations

Any equation in x which can be written in the form

$$ax^2 + bx + c = 0, \quad a \neq 0$$

with the highest power being x^2, is called a **quadratic equation** or quadratic for short. It is an equation of the second degree. Quadratic equations have two solutions or roots, which are sometimes equal.

The starting point *always* when solving a quadratic equation is to put everything on the left-hand side of the equation and to have zero on the right-hand side, so that you have an equation of the form $ax^2 + bx + c = 0$. When you have done this there are three common ways of solving the quadratic equation: by factorisation, by completing the square and by using a standard formula.

Solving quadratic equations by factorisation

This is usually the quickest way of solving a quadratic. (Page 26 shows you how to find out whether a quadratic will factorise or not.)

After moving all the terms to the left-hand side of the equation you factorise the left-hand side.

Example 9

Solve $x^2 - 4 = 0$.

$$x^2 - 4 = 0$$

Factorise the left-hand side:

$$(x - 2)(x + 2) = 0$$

So either $\qquad x - 2 = 0 \text{ and } x = 2$

or $\qquad x + 2 = 0 \text{ and } x = -2$

Example 10

Solve $3x^2 + 4x = 0$.

$$3x^2 + 4x = 0$$

So: $\qquad x(3x + 4) = 0$

Either $\qquad x = 0$

or $\qquad 3x + 4 = 0 \text{ and } x = -\frac{4}{3} = -1\frac{1}{3}$

Example 11

Solve $x^2 - 5x - 14 = 0$.

$$x^2 - 5x - 14 = 0$$

So: $\qquad (x + 2)(x - 7) = 0$

Either $\qquad x + 2 = 0 \text{ and } x = -2$

or $\qquad x - 7 = 0 \text{ and } x = 7$

Example 12

Solve $2x^2 - x - 3 = 0$.

$$2x^2 - x - 3 = 0$$

So: $\qquad (2x - 3)(x + 1) = 0$

Either $\qquad 2x - 3 = 0 \text{ and } x = 1\frac{1}{2}$

or $\qquad x + 1 = 0 \text{ and } x = -1$

Solving quadratic equations by completing the square

The idea of this method of solving a quadratic is to try to get a perfect square on the left-hand side of the equation so that, by taking the square root of both sides of the equation, the roots of the equation can be found. For this method you have to use the identities

$$x^2 + 2ax + a^2 \equiv (x + a)^2$$

and $$x^2 - 2ax + a^2 \equiv (x - a)^2$$

The next two examples show you how this is done. Use this method when factorisation is difficult or impossible.

Example 13

Solve $x^2 + 5x - 3 = 0$, giving the roots to 2 decimal places.

$$x^2 + 5x - 3 = 0$$

Take the constant to the right-hand side:

$$x^2 + 5x = 3$$

Compare the left-hand side of this with the left-hand side of the first of the two identities given, $x^2 + 2ax + a^2 \equiv (x + a)^2$. You see that

$$2ax = 5x$$

So: $$a = \tfrac{5}{2} \text{ and } a^2 = \left(\tfrac{5}{2}\right)^2$$

You can complete the square by adding $\left(\tfrac{5}{2}\right)^2$ to $x^2 + 5x$:

$$x^2 + 5x + \left(\tfrac{5}{2}\right)^2 \equiv \left(x + \tfrac{5}{2}\right)^2$$

You have added $\left(\tfrac{5}{2}\right)^2$ to the left-hand side of the original equation, so you must also add it to the right-hand side:

$$x^2 + 5x + \left(\tfrac{5}{2}\right)^2 = 3 + \left(\tfrac{5}{2}\right)^2$$

$$\left(x + \tfrac{5}{2}\right)^2 = 3 + \left(\tfrac{5}{2}\right)^2 = 3 + \tfrac{25}{4} = \tfrac{37}{4}$$

Taking the square root:

$$x + \tfrac{5}{2} = \pm 3.041$$

Either $$x + 2.5 = 3.041 \text{ and so } x = 0.54 \,(2\,\text{d.p.})$$

or $$x + 2.5 = -3.041 \text{ and so } x = -5.54 \,(2\,\text{d.p.})$$

The steps to follow when using this method are:

step 1 Divide the equation by the coefficient of x^2.

step 2 Take the constant to the right-hand side.

step 3 Complete the square.

step 4 Find the square root.

step 5 Make x the subject.

Example 14

Solve $2x^2 - 8x - 5 = 0$ giving the answers to 2 decimal places.

$$2x^2 - 8x - 5 = 0$$

step 1 :
$$x^2 - 4x - \frac{5}{2} = 0$$

step 2 :
$$x^2 - 4x = \frac{5}{2}$$

step 3 :
$$(x - 2)^2 - \frac{5}{2} + 2^2 = 6.5$$

step 4 :
$$x - 2 = \pm 2.550$$

step 5 :
$$x - 2 = 2.550 \text{ and so } x = 4.55 \,(2\,\text{d.p.})$$

or
$$x - 2 = -2.550 \text{ and so } x = -0.55 \,(2\,\text{d.p.})$$

Many people find the method of completing the square rather difficult so it has been generalised to give a formula. If you memorise this formula and use it correctly you will be able to solve *all* quadratic equations.

Solving quadratic equations using the quadratic formula

Start with the quadratic equation $ax^2 + bx + c = 0$ and solve it by completing the square.

$$ax^2 + bx + c = 0$$

step 1 :
$$x^2 + \frac{b}{a}x + \frac{c}{a} = 0 \qquad \text{Divide through by } a$$

step 2 :
$$x^2 + \frac{b}{a}x = \frac{-c}{a} \qquad \text{make } \frac{-c}{a} \text{ to other side}$$

step 3 :
$$\left(x + \frac{b}{2a}\right)^2 = \frac{-c}{a} + \left(\frac{b}{2a}\right)^2$$

Complete the square
Denominators ×2 and $\left(\frac{b}{2a}\right)^2$
to other side

step 4 :
$$x + \frac{b}{2a} = \pm \sqrt{\left[\frac{-c}{a} + \left(\frac{b}{2a}\right)^2\right]}$$

$\sqrt{}$ *while she bang*

step 5 :
$$x = \frac{-b}{2a} \pm \sqrt{\left[\frac{-c}{a} + \left(\frac{b}{2a}\right)^2\right]}$$

move $\frac{b}{2a}$ *to other side*

That is :
$$x = \frac{-b}{2a} \pm \sqrt{\left(\frac{-c}{a} + \frac{b^2}{4a^2}\right)}$$

$\left(\frac{b}{2a}\right)^2 \Rightarrow \frac{b^2}{4a^2}$

$$= \frac{-b}{2a} \pm \sqrt{\left(\frac{-4ac + b^2}{4a^2}\right)}$$

bd the bracket fractions together

so
$$x = \frac{-b \pm \sqrt{(b^2 - 4ac)}}{2a}$$

taken out 2A

This is known as the quadratic formula. You should memorise it.

Example 15

Solve the equation $5x^2 - 11x + 4 = 0$ giving the roots to 2 decimal places. This is the general equation $ax^2 + bx + c = 0$ with $a = 5$, $b = -11$ and $c = 4$.

Substituting these values into the formula:

$$x = \frac{-b \pm \sqrt{(b^2 - 4ac)}}{2a}$$

$$= \frac{11 \pm \sqrt{[121 - (4 \times 5 \times 4)]}}{10}$$

$$= \frac{11 \pm \sqrt{41}}{10}$$

$$= \frac{11 \pm 6.403}{10}$$

$$= 1.74 \,(2 \,\text{d.p.}) \,\text{or} \,0.46 \,(2 \,\text{d.p.})$$

How to check whether a quadratic can be factorised

The expression $(b^2 - 4ac)$ is called the **discriminant. If the discriminant has an exact square root then the quadratic equation will factorise into two linear factors**. You will save yourself a lot of time and energy if you use the method of factorisation whenever you can.

Exercise 2B

Solve the following quadratic equations by the method of factorisation:

1	$x^2 - 49 = 0$	**2**	$4x^2 - 25 = 0$
3	$x^2 = 144$	**4**	$x^2 + 5x = 0$
5	$x^2 - 7x = 0$	**6**	$x^2 - x - 2 = 0$
7	$x^2 + x - 6 = 0$	**8**	$x^2 + 7x + 10 = 0$
9	$x^2 - 9x + 20 = 0$	**10**	$x^2 + 6x + 9 = 0$
11	$x^2 + 12x + 32 = 0$	**12**	$x^2 - 9x - 10 = 0$
13	$x^2 - x - 12 = 0$	**14**	$x^2 + 4x - 12 = 0$
15	$x^2 \quad 11x \quad 12 = 0$	**16**	$x^2 + 13x + 12 - 0$
17	$2x^2 - 3x - 2 = 0$	**18**	$2x^2 - x - 1 = 0$
19	$2x^2 + 5x + 2 = 0$	**20**	$3x^2 - 7x + 2 = 0$
21	$3x^2 + 19x - 14 = 0$	**22**	$3x^2 - 13x - 10 = 0$
23	$12x^2 - 7x + 1 = 0$	**24**	$6x^2 - 5x - 6 = 0$
25	$8x^2 - 22x + 15 = 0$	**26**	$28x^2 - 51x - 27 = 0$
27	$15x^2 + 22x + 8 = 0$	**28**	$20x^2 - 31x - 9 = 0$

Solve the following quadratic equations by (i) completing the square, (ii) using the formula, giving your answers to 2 decimal places:

31	$x^2 - 2x - 1 = 0$	**32**	$x^2 + 8x + 6 = 0$
33	$x^2 - 3x - 5 = 0$	**34**	$10x^2 - 2x - 3 = 0$
35	$7x^2 + 6x + 1 = 0$	**36**	$4x^2 - 3x - 2 = 0$
37	$3x^2 - 5x + 1 = 0$	**38**	$5x^2 + 3x - 3 = 0$
39	$3x^2 - 4x = 2$	**40**	$(2x - 3)^2 = 2x$

2.4 Solving two simultaneous equations, one linear and one quadratic

When you can solve two linear simultaneous equations and can solve quadratic equations, you can also solve two simultaneous equations of which one is linear and the other quadratic. To solve these make either x or y the subject of the linear equation, substitute this into the quadratic equation and then solve the quadratic equation you are left with.

Notice that two simultaneous equations, one linear and one quadratic, have, in general, two pairs of solutions. Each pair involves a value of x and a value of y.

Example 16

Solve these simultaneous equations:

$$x - 2y = 7$$
$$x^2 + 4y^2 = 37$$

Taking the linear equation and making x the subject gives $x = 7 + 2y$.

Substituting into the second equation gives

$$(7 + 2y)^2 + 4y^2 = 37$$
$$49 + 28y + 4y^2 + 4y^2 - 37 = 0$$
$$8y^2 + 28y + 12 = 0$$

Divide by 4: $\qquad\qquad 2y^2 + 7y + 3 = 0$

Solve the quadratic: $\qquad (2y + 1)(y + 3) = 0$

So either $\qquad\qquad 2y + 1 = 0 \Rightarrow y = -\frac{1}{2}$

or $\qquad\qquad\qquad y + 3 = 0 \Rightarrow y = -3$

Now substitute these values of y into the linear equation to find the corresponding values of x.

When $y = -\frac{1}{2}$, $x = 7 - 1 = 6$

When $y = -3$, $x = 7 - 6 = 1$

So the solutions are:

$$\left\{ \begin{matrix} x = 6 \\ y = -\frac{1}{2} \end{matrix} \right\} \text{and} \left\{ \begin{matrix} x = 1 \\ y = -3 \end{matrix} \right\}$$

Notice that to specify the solutions correctly you need to show which value of x is paired with which value of y. An easy way of doing this is by using braces { }. Alternatively, the solutions can be written in brackets, for example: $(6, -\frac{1}{2})$, $(1, -3)$.

Exercise 2C

Solve the simultaneous equations:

1 $x + y = 1$
   $16x^2 + y^2 = 65$

2 $2x + y = 1$
   $x^2 + y^2 = 1$

3 $y - x = 2$
   $2x^2 + 3xy + y^2 = 8$

4 $x - 2y = 7$
   $x^2 + 4y^2 = 37$

5 $x + y = 9$
   $x^2 - 3xy + 2y^2 = 0$

6 $x = 2y$
   $x^2 + 3xy = 10$

7 $y - x^2 + 3$
   $y = 4x$

8 $u \quad v = 3$
   $u^2 + v^2 = 89$

9 $x + 2y = -3$
   $x^2 - 2x + 3y^2 = 11$

10 $y - x = 4$
   $2x^2 + xy + y^2 = 8$

2.5 Solving linear and quadratic inequalities in one variable

Linear inequalities are similar to linear equations except that the $=$ sign in the equation is replaced by one of four inequality signs:

$>$ is greater than

$\geqslant$ is greater than or equal to

$<$ is less than

$\leqslant$ is less than or equal to

So 'four is greater than 1' can be written $4 > 1$, and 'minus 3 is less than 10' can be written $-3 < 10$, and so on.

The rules for manipulating linear inequalities are, with one major exception, the same as those used for manipulating linear equations. So if $x > y$ then

$$x \pm c > y \pm c$$

and
$$mx > my \text{ if } m \text{ is positive}$$

and
$$\frac{x}{m} > \frac{y}{m} \text{ if } m \text{ is positive}$$

The one major exception comes if you try to multiply or divide both sides of a linear inequality by a negative number. In these cases the inequality sign has to be *reversed*. That is, if $x > y$ then

$$nx < ny \text{ if } n \text{ is negative}$$

and
$$\frac{x}{n} < \frac{y}{n} \text{ if } n \text{ is negative}$$

The validity of this rule can easily be seen from two examples:

(i) $7 > -4$, i.e. 'seven is greater than minus four'.

But if you multiply by -2, then $-14 < 8$, i.e. 'minus fourteen is less than eight'.

(ii) $-15 < 3$, i.e. 'minus fifteen is less than three'.

But if you divide by -3 then $5 > -1$, i.e. 'five is greater than minus one'.

Example 17
Find the set of values for which $8 + 3x \geqslant 23$.

$$8 + 3x \geqslant 23$$
$$3x \geqslant 23 - 8$$
$$3x \geqslant 15$$
$$x \geqslant 5$$

You can show this solution diagrammatically on a number line:

The heavy dot at 5 indicates that 5 is included in the solution.

Example 18
Find the set of values for which $15 - 7x < 24 - 4x$.

$$15 - 7x < 24 - 4x$$
$$-7x + 4x < 24 - 15$$
$$-3x < 9$$
$$3x > -9$$
$$x > -3$$

This solution is shown on the number line like this:

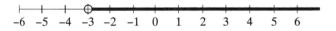

In this case, the circle at -3 indicates that -3 is *not* included in the solution.

You can also manipulate **quadratic inequalities** to find the set of values for which they are true. However, manipulating a quadratic inequality is somewhat more complex than manipulating a linear inequality. You must examine either the graph of the quadratic function or the signs of the factors of the quadratic function.

Example 19

Find the set of values for which $x^2 - 5x + 4 > 0$.

$$x^2 - 5x + 4 > 0$$

factorises to:
$$(x - 4)(x - 1) > 0$$

A sketch of the quadratic function $y = x^2 - 5x + 4$ looks like this:

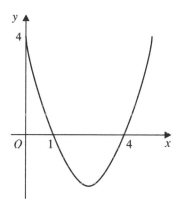

From the graph you can see that $(x^2 - 5x + 4)$ is greater than zero (i.e. it lies above the x-axis) if $x < 1$ or $x > 4$ and so this is the solution.

The alternative way of obtaining the solution is to look at the signs of the factors in the regions into which the graph is divided by the critical values. The **critical values** are the values of x where the graph cuts the x-axis; that is, they are the roots of the quadratic equation $(x - 4)(x - 1) = 0$. Since they are the roots of the corresponding quadratic equation they can be found, if necessary, without sketching the graph of the quadratic function.

Draw up a table which gives the *sign* of the factors in each of the three regions:

x	$x < 1$	$1 < x < 4$	$x > 4$
$x - 4$	$-ive$	$-ive$	$+ive$
$x - 1$	$-ive$	$+ive$	$+ive$
$(x - 4)(x - 1)$	$+ive$	$-ive$	$+ive$

The table shows that, once again, $(x - 4)(x - 1)$ is positive if $x < 1$ or $x > 4$.

Example 20

Find the set of values for which $2x^2 - 7x - 15 \leqslant 0$.

$$2x^2 - 7x - 15 \leqslant 0$$
$$(2x + 3)(x - 5) \leqslant 0$$

The critical values are thus $-\frac{3}{2}$ and 5, and so the three regions are $x \leqslant -\frac{3}{2}$, $-\frac{3}{2} \leqslant x \leqslant 5$, $x \geqslant 5$.

x	$x \leqslant -\frac{3}{2}$	$-\frac{3}{2} \leqslant x \leqslant 5$	$x \geqslant 5$
$2x + 3$	$-ive$	$+ive$	$+ive$
$x - 5$	$-ive$	$-ive$	$+ive$
$(2x + 3)(x - 5)$	$+ive$	$-ive$	$+ive$

From the table you can see that $(2x + 3)(x - 5) \leqslant 0$ for all values of x in the set $-\frac{3}{2} \leqslant x \leqslant 5$.

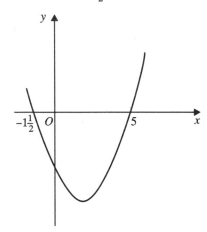

The sketch of the curve $y = (2x + 3)(x - 5)$ confirms that the function is negative or zero (i.e. lies below or crosses the x-axis) for $-1\frac{1}{2} \leqslant x \leqslant 5$.

Exercise 2D

Find the set of values for which:

1 $3x + 4 > 7$

2 $7x - 2 < 19$

3 $3 > 4x + 11$

4 $26 < 8 - 9x$

5 $6x - 7 + 2x - 9 \geqslant 0$

6 $9x - 17 + 6x - 8 \leqslant 20$

7 $5x - 30 < x - 4$

8 $17 - 3x \geqslant 9x + 45$

9 $4 + 5x < 8 - 11x$

10 $7x - 5 + 4x \geqslant 20 - 6x + 26$

11 $2(3x - 7) + 3 > 13 - 2x$

12 $3(2x - 7) > 5(6 - x) + 4$

13 $2(x-7)+4(3-2x) \leqslant -26$

14 $3(x-7)+2(2x-9) < 2(12-x)$

15 $8(2x-4)-9x \leqslant 3$

16 $2(3x-7)+3(x+1) \geqslant 3(10-x)+1$

17 $-7-4(x-3) > 23+2x$

18 $5(2x+7)+20 \leqslant 3(x+9)$ 1

19 $3(5x+7)+2(x-9) < 10(6-x)+3(1-3x)$

20 $-15+7(2x+3) \geqslant 6x+8(2-x)$

Find the set of values for which:

21	$x^2+7x+10 \geqslant 0$	**22**	$x^2-5x+6 \leqslant 0$
23	$x^2+x-12 > 0$	**24**	$x^2-9x+18 < 0$
25	$x^2-7x-18 \leqslant 0$	**26**	$x^2+11x+28 > 0$
27	$2x^2-11x+12 > 0$	**28**	$3x^2+2x-8 > 0$
29	$3x^2-19x-14 < 0$	**30**	$2x^2-13x+21 > 0$
31	$2x^2+13x+20 \leqslant 0$	**32**	$5x^2-4x-9 > 0$
33	$3x^2+34x+63 \geqslant 0$	**34**	$4x^2-23x+15 < 0$
35	$3x^2-23x+30 > 0$	**36**	$3x^2-4x-2 \leqslant 2$
37	$2x^2+x-6 < 0$	**38**	$12x^2-5x-2 \leqslant 0$
39	$21x^2+5x-6 \geqslant 0$	**40**	$20+7x-6x^2 > 0$

SUMMARY OF KEY POINTS

1 A linear equation is one of the form $ax+b=0$. It is sometimes called an equation of the first degree.

2 In general, two linear equations of the form $ax+by+c=0$ in the same two unknowns x and y can be solved simultaneously. The solution of these equations is a pair of values – a value of x and a value of y. (This is true *unless* the second equation is of the form $max+mby+d=0$ in which case there is not a unique solution.)

3 There are three methods of solving simultaneous linear equations in two unknowns – graphical methods, substitution and elimination.

4 Any equation that can be written in the form $ax^2+bx+c=0$, $a \neq 0$, is called a quadratic equation, or an equation of the second degree.

5   There are three methods of solving quadratic equations: factorisation, completing the square and formula.

6   The formula for solving $ax^2 + bx + c = 0$ is

$$x = \frac{-b \pm \sqrt{(b^2 - 4ac)}}{2a}$$

7   To solve two simultaneous equations in two unknowns, one linear and one quadratic, you make x or y the subject of the linear equation, substitute it into the quadratic equation, and then solve the quadratic equation you are left with.

8   A linear inequality in one unknown can be solved using the same rules as for solving a linear equation to obtain the critical value. Remember that if you multiply or divide the inequality by a negative number, the inequality sign changes.

9   The solution of a quadratic inequality depends on finding the critical values, either by solving a quadratic equation or by considering the graph of the function.

Functions

3.1 Mappings and functions

In mathematics you will encounter many situations in which each member of one set of numbers is related to a member of another set of numbers by some well understood rule. For example, the rule $y = 3x + 2$ relates the set of numbers $x = 1$, 2, 3 to the set of numbers $y = 5$, 8, 11.

This section looks at different ways of representing relationships between sets of numbers. Some methods make relationships more visual and easier to remember; they are all intended to help make it easier to record and understand relationships.

Introducing the notation of functions

First think about two sets of numbers X and Y related by a rule. The first set X has members x_1, x_2, x_3, ... which produce the members y_1, y_2, y_3, ... of the set Y when you apply the rule. One way of recording the relationship is by grouping the members in ordered pairs

$$(x_1, y_1), (x_2, y_2), (x_3, y_3), \ldots$$

Notice that each x value produces just one y value. The rule together with the set X on which it operates is called a **function**. It is usually symbolised by a letter such as f. The set X is called the **domain** of f: here domain means the set of numbers which the function is operating on. The set Y is called the **range** of f: meaning the range of numbers produced by the function.

In symbols we write the function like this:

$$f : x \mapsto y, \quad x \in X, y \in Y$$

This means: 'the function f maps each element x of the set X onto an element y of the set Y'.

$\in$ is shorthand for 'is a member of' and $x \in X$ is shorthand for 'x is a member of the set X'. Sometimes you will work with numbers from the following sets so you need to learn their symbols:

$\mathbb{Z}$ the set of integers	$\mathbb{N}$ the set of natural numbers
$\mathbb{Q}$ the set of rational numbers	$\mathbb{R}$ the set of real numbers

For example, $x \in \mathbb{R}$ is shorthand for 'x is a member of the set of real numbers' or 'x is defined over the real numbers'. The domain of a function is often given in this way.

The function f is sometimes shown as a mapping diagram in which each member of set X is mapped to a member of set Y by an arrow:

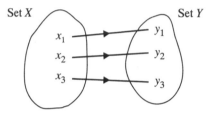

Graphing a function

More often f is shown as a graph on which the 'points' (x, y) are plotted. The domain is all the x values and the range is all the y values. The x values are sometimes called the **independent variable**. The y values are then called the **dependent variable** – they depend on the x values and on the function.

The equation of the resulting line or curve can be written in the form $y = f(x)$. In this notation $f(x)$ is called the image of x under the function f.

The following examples illustrate the ideas, language and notation of functions.

Example 1
A function f is defined by

$$f : x \mapsto 3x - 1, \quad x \in \mathbb{R}, -1 \leqslant x \leqslant 4$$

This means the rule of f is 'multiply each value of x by 3 and take away 1'. This could be shown in a flow chart by

The domain of f is all real values from -1 to 4 inclusive. When $x = -1$, $f(x) = 3 \times (-1) - 1 = -4$ and we write $f(-1) = -4$.

Similarly

$$f(4) = (3 \times 4) - 1 = 11$$

You can find other values the same way, for example: $f(0) = -1$, $f(1) = 2$ and $f(2) = 5$. Here is the graph of f:

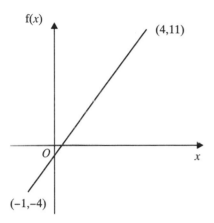

You can see that it is a line segment joining the points $(-1, -4)$ and $(4,11)$.

The range of f is written as $y \in \mathbb{R}$, $-4 \leqslant y \leqslant 11$.

Example 2

Find the values of $f(-2)$, $f(1)$ and $f(0.3)$ for the function f given by:

$$f : x \mapsto x^3 + 3x, \ x \in \mathbb{R}$$

Take a value of x, cube it, then add 3 times the value of x taken to obtain the corresponding value of $f(x)$. For example:

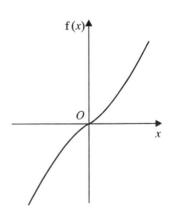

$f(-2) = (-2)^3 + 3(-2) = -8 - 6 = -14$

$f(1) = (1)^3 + 3(1) = 1 + 3 = 4$

$f(0.3) = (0.3)^3 + 3(0.3) = 0.027 + 0.9 = 0.927$

The rule of this function cannot easily be represented by a flow chart but it is worth noting that $f(x) = -f(-x)$ and this is helpful in sketching the graph of f.

The range of f is all real numbers, written as $y \in \mathbb{R}$.

Example 3

Find the range of the function f, defined by

$$f : x \mapsto \frac{1}{x}, \ x \in \mathbb{R}, \ x \neq 0$$

We have to exclude zero from the domain of f because $\frac{1}{x}$ is undefined in value at $x = 0$. As you can see from the graph of f, all other real values are contained in the range of f. The range of f is therefore $y \in \mathbb{R}, \ y \neq 0$.

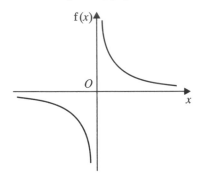

Exercise 3A

In questions 1–5, f is a function whose domain is the set of real numbers. Within the interval given, sketch the graph of f and determine the range of f.

1   $f : x \mapsto 2x - 3, \quad 0 \leqslant x \leqslant 6$

2   $f : x \mapsto 4 - x, \quad -2 \leqslant x \leqslant 2$

3   $f : x \mapsto x^2, \quad -4 \leqslant x \leqslant 4$

4   $f : x \mapsto \frac{1}{x}, \quad 2 \leqslant x \leqslant 10$

5   $f : x \mapsto 2^x, \quad -2 \leqslant x \leqslant 4$

6   Given that $f(x) = \frac{1}{x^2 + 1}$, find the values of

   (a) $f(-2)$ (b) $f(-1)$ (c) $f(0)$.

   Sketch the graph of the curve with equation $y = f(x)$ for all real values of x in the interval $-2 \leqslant x \leqslant 2$.

7   Find the range of the function f which is defined by

$$f : x \mapsto \frac{1}{x + 1}, \ x \in \mathbb{R}, \ 0 \leqslant x \leqslant 7$$

   by first sketching the graph of f.

In each of the following functions, the rule and the range are given. By sketching the graph, find the domain.

8 $f : x \mapsto 3x + 4,$ range $1 \leqslant y \leqslant 12$

9 $g : x \mapsto 5 - 2x,$ range $-7 \leqslant y \leqslant 11$

10 $h : x \mapsto 3^x,$ range $1 \leqslant y \leqslant 729$

3.2 Composite functions

Most functions which you will meet are combinations of two or more basic functions. These are called **composite functions**

Example 4

Discuss the function f where x is defined over the real numbers by

$$f : x \mapsto 2x - 3$$

The function f is a combination of the two simpler functions g and h which are given by

$g : x \mapsto 2x$ (the 'doubling' function

$h : x \mapsto x - 3$ (the 'take away 3' function)

The diagram shows how you can combine the functions g and h to obtain the same mapping as given by the function f.

The diagram makes it quite clear that g must be applied *first* and h applied *second* in order to obtain f. We write $h[g(x)]$ when we want g to precede h because the rule of h is applied to the result $g(x)$, which has been obtained already by applying the rule of function g.

$h[g(x)]$ is usually abbreviated to $hg(x)$. *Remember* that hg means '*do g first and then h*'. If the order of applying g and h is reversed, then a different function e is formed.

The function e is given by $e : x \mapsto 2x - 6$.

Example 5

The functions p and q are defined over the real numbers by

$$p : x \mapsto x^2$$

$$q : x \mapsto 5 - 4x$$

Find the values of (a) pq(3)   (b) qp(3)   (c) qq(3).

(a) Note that $q(3) = 5 - 4 \times 3 = 5 - 12 = -7$.

This gives $pq(3) = p[q(3)] = p(-7) = (-7)^2 = 49$

(b) You now have $p(3) = 3^2 = 9$.
This gives $qp(3) = q[p(3)] = q(9) = 5 - 4 \times 9 = -31$

(c) Using $q(3) = -7$ from (a),
$q[q(3)] = q(-7) = 5 - 4(-7) = 5 + 28 = 33$

Exercise 3B

1   The functions f, g and h each have the set of real numbers as
their domain and are defined by

f:$x \mapsto 7 - 2x$

g:$x \mapsto 4x - 1$

h:$x \mapsto 3(x - 1)$

(a) Find the following composite functions in terms of x.
(i) fg (ii) gf (iii) fh (iv) hf (v) gh (vi) hg
(b) Evaluate (i) fg(5) (ii) ff(−2) (iii) fh(0) (iv) hg($-\frac{1}{2}$)
(v) fgh(3) (vi) ghf(3)
(c) Find the values of x for which (i) $fg(x) = -15$
(ii) $gh(x) = 11$ (iii) $hgf(x) = 102$

2   The function g has as its domain all nonzero real numbers
and is given by

$$g(x) = x - \frac{1}{x}$$

Find the values of (a) g(2)   (b) gg(2)   (c) g($-\frac{1}{3}$).
Find also the values of x for which $g(x) = \frac{15}{4}$.

3   Functions g and h whose domains are the set of real numbers
are defined by

$$g : x \mapsto 3 - x \qquad h : x \mapsto x^2 - 1$$

Find the values of (a) gh(2) (b) hh(2) (c) hg(−3) (d) ggg(0).

4   The composite functions f and g, given by $f(x) = x^3 + 2$ and $g(x) = \dfrac{1}{x} - 3$, could each be formed from two simpler functions. Write down these functions. Assuming that f and g have the set of real numbers as their domain, find the values of fg(2) and gf(2).

5   Given that $f(x) = ax + b$ and $g(x) = bx + a$, and that a and b are constant and unequal, find a relation between a and b for which

$$fg(x) = gf(x)$$

6   The functions f and g are defined over the real numbers and

$$f : x \mapsto 1 + \frac{x}{2} \qquad g : x \mapsto x^2$$

Form the composite functions (a) fg (b) gf (c) ff, expressing each in terms of x in a simplified form.

7   The functions p and q are defined by

$$p : x \mapsto Ax + B \qquad q : x \mapsto Cx + D$$

where A, B, C and D are constants. Given that $pq(x) = qp(x)$, show that

$$D(A - 1) = B(C - 1)$$

8   The functions f and g are defined by

$$f : x \mapsto 2x^2 + 1 \qquad g : x \mapsto 3x - 2$$

Find the functions, fg, gf, ff and gg in terms of x, and sketch their graphs.

3.3 Inverse functions

For any function each member of the domain corresponds to one, and only one, member of the range. Sometimes it is possible to choose a domain and a range so that, if $f : x \mapsto y$ is a function, then $y \mapsto x$ is also a function. We call this function the **inverse function** of f, written

$$f^{-1} : y \mapsto x$$

An inverse function f^{-1} can exist if, and only if, the function f is a one–one mapping. That is, each member of the domain must be paired with one and only one member of the range and each member of the range is paired with only one member of the domain.

Example 6

Discuss the function f given by $f(x) = x^2$.

The function f given by $f(x) = x^2$ with domain the set of real numbers is not one–one. This is clear because, for example, both the numbers 2 and -2 in the domain map to 4 in the range. This type of function is known as a many–one mapping. However, it is worth noting that if you restrict the domain to the set of positive real numbers, you turn the function x^2 into a one–one mapping.

Starting with the one–one function f defined over a suitable domain, you can find the inverse function f^{-1} by following these steps, starting from $y = f(x)$.

1. Interchange x and y to give $x = f(y)$.

2. Express y in terms of x to give $y = f^{-1}(x)$.

3. The inverse function is f^{-1} with domain the same as the range of f.

Example 7

The function f is defined by

$$f : x \mapsto 3x - 1, \qquad x \in \mathbb{R}, \ -1 \leqslant x \leqslant 4$$

Find the inverse function f^{-1} and sketch the graphs of the functions f and f^{-1} on the same axes.

First take $y = 3x - 1$ and change over x and y to obtain

$$x = 3y - 1$$

Rearranging, you get

$$3y = x + 1 \quad \text{and} \quad y = \frac{x + 1}{3}$$

That is,

$$f^{-1}(x) = \frac{x + 1}{3}$$

Referring to Example 1, where we discussed the same function, you know that the range of f is $-4 \leqslant x \leqslant 11$. This is also the domain of f^{-1}.

Then, $f^{-1} : x \mapsto \dfrac{x + 1}{3}, \qquad x \in \mathbb{R}, \quad -4 \leqslant x \leqslant 11$.

Finally, draw the graphs of f and f^{-1} on the same axes. Notice that the graphs of f and f^{-1} are reflections of each other in the line $y = x$.

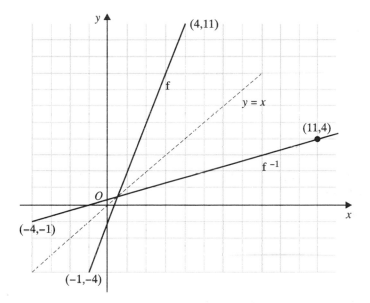

This is true for all one–one functions and their inverse functions because

$$ff^{-1}(x) = x = f^{-1}f(x)$$

Example 8

Sketch the graph of the function f defined by

$$f : x \mapsto (x - 2)^2, \qquad x \in \mathbb{R}$$

Explain why f cannot have an inverse function. By making an appropriate change to the domain of f define a function g with the same rule as f which has an inverse function g^{-1}. Express g^{-1} in terms of x.

The sketch of a graph should not only indicate the general shape but should also include the coordinates of any points at which the curve meets the coordinate axes, when these can be easily calculated.

The curve with equation $y = (x - 2)^2$ is a parabola of U shape with its vertex, the lowest point, at the point $A(2,0)$, as shown. This is also the sketch of f. This sketch shows that f is a many–one mapping because, for example, both 0 and 4 in the domain map to 4 in the range. Therefore, the function f cannot possess an inverse function.

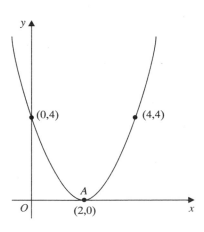

Now consider the function g defined by

$$g : x \mapsto (x - 2)^2, \qquad x \in \mathbb{R}, x \geqslant 2$$

By restricting the domain of g, compared with that of f, we have made g a one–one function, as shown in the next diagram.

To find the inverse function g^{-1} of g, write

$$y = (x - 2)^2$$

Interchange x and y : $x = (y - 2)^2$

Expressing y in
terms of x: $y - 2 = \pm\sqrt{x}$

That is, $y = 2 + \sqrt{x}$

Note that we choose the $+$ sign on the square root so that the graph
of g^{-1} is the reflection of the graph of g in the line $y = x$.

The inverse function g^{-1} is therefore given by

$$g^{-1} : x \mapsto 2 + \sqrt{x}, \quad x \in \mathbb{R}, x \geqslant 0$$

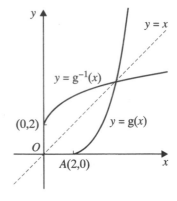

Exercise 3C

In all questions, the domain of each function is the set of real
numbers unless specifically stated otherwise. In questions 1–5,
sketch the given function f. Find f^{-1}. Sketch on the same diagram
the functions f and f^{-1}.

1 $f(x) = 2x$ **2** $f(x) = 2x + 3$ **3** $f(x) = 1 - 2x$

4 $f(x) = x^2, x \leqslant 0$ **5** $f(x) = \dfrac{1}{x}, x > 0$

In questions 6–11 some simple functions are given, where p and q
are positive constants. Find the inverse function for each.

6 $f(x) = x + p$ **7** $f(x) = px$ **8** $f(x) = p - x$

9 $f(x) = px + q$ **10** $f(x) = \dfrac{p}{x}, x \neq 0$

11 $f(x) = \dfrac{1}{(px + q)}, x \neq \dfrac{-q}{p}$

In questions 12–15, find the range of the function whose domain
is the set of positive real numbers. Find also the inverse function.

12 $f(x) = \dfrac{1}{x^2}$ **13** $f(x) = \dfrac{1}{x + 1}$

14 $f(x) = (x + 1)^2 - 1$ **15** $f(x) = x^2 + 4x + 5$

16 Find the values of A and B so that

$$(x - A)^2 + B \equiv x^2 - 10x + 29$$

for all values of x.

The function $f : x \mapsto x^2 - 10x + 29, x \in \mathbb{R}, x \geqslant k$ is one–one. Find the smallest possible positive value of k and the range of f in this case. Find also the inverse function when k takes its smallest possible positive value.

17 Find the inverse function g^{-1} of $g(x) = \dfrac{1}{x - 4}, x \neq 4.$

18 Sketch the function 2^x taking values of x in the interval $-2 \leqslant x \leqslant 3$ as domain. Sketch the graph of the inverse function, stating its domain.

3.4 The modulus function $|x|, x \in \mathbb{R}$

The **modulus** of any number, positive or negative, is the size of the number without a sign attached.

For example, the modulus of 5, written $|5|$, is 5 and the modulus of -3, written $|-3|$, is 3.

The modulus function, written as $|x|$, is defined as

$$|x| = \begin{cases} x \text{ for all real } x \geqslant 0 \\ -x \text{ for all real } x < 0 \end{cases}$$

Here is the graph of $y = |x|$. It consists of the line $y = x$ for $x \geqslant 0$ and the line $y = -x$ for $x < 0$.

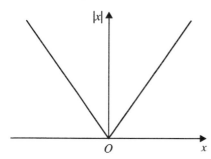

The statement $|x| < p$ means that x is *numerically* less than p; that is $-p < x < p$, where p is a fixed positive number.

3.5 Sketching graphs

Here are sketch graphs of $y = f(x)$ for some simple standard functions such as $f(x) = x^n$, for $n = 1, 2, 3, -1$ and $\frac{1}{2}$.

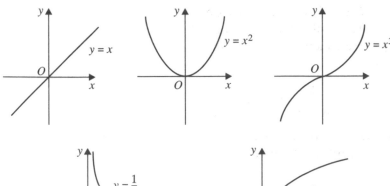

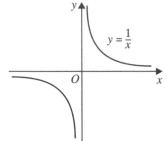
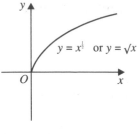

Even and odd functions

An even function has the property $f(x) = f(-x)$. The graph of the function is symmetrical about the line $x = 0$ (the y-axis). If (a, b) is on the graph, so is $(-a, b)$.

$f(x) = x^2$ is an even function (see above).

An odd function has the property $f(x) = -f(-x)$, which is more often given as $f(-x) = -f(x)$. If (a, b) is on the graph, so is $(-a, -b)$.

$f(x) = \dfrac{1}{x}$, $f(x) = x^3$, $f(x) = x$ are odd functions (see above).

3.6 Transforming graphs

Let's look at the effect of a few simple transformations on the graph of $y = f(x)$.

The transformation $y = af(x)$

All points on the x-axis remain unchanged under this transformation.

The general point $[t, f(t)]$ maps to the point $[t, af(t)]$, where t is any point in the domain of f. The following example shows the effect of taking different values of a.

Example 9

Starting with the basic curve $y = x^2$, sketch the curve $y = ax^2$ when $a = 2, \frac{1}{2}, -1, -2$ and $-\frac{1}{2}$.

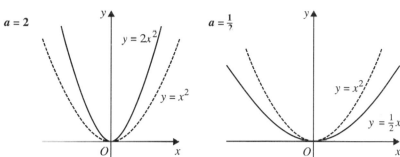

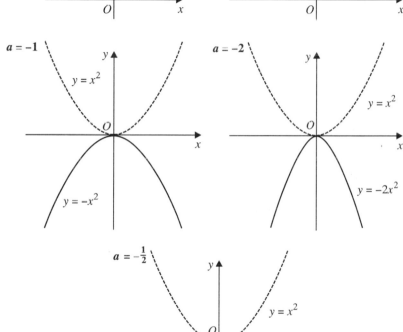

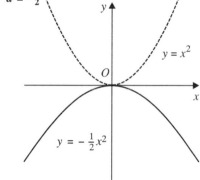

The transformation $y = f(x) + a$

If $a > 0$, the transformation translates the graph a steps in the positive y direction.

If $a < 0$, the transformation translates the graph a steps in the negative y direction.

Example 10

Starting from the basic curve $y = \dfrac{1}{x}$, sketch the curves $y = \dfrac{1}{x} + 3$ and $y = \dfrac{1}{x} - 3$.

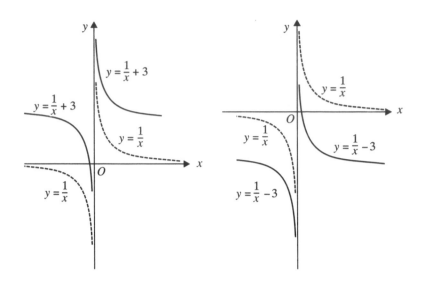

The transformation $y = f(x + a)$

If $a > 0$, the transformation translates the graph a steps to the left, that is, in the negative x direction.

If $a < 0$, the graph is translated a steps to the right, that is, in the positive x direction.

Example 11

Starting from the graph of the function $|x|$, sketch the graph of the functions $|x + 3|$ and $|x - 4|$.

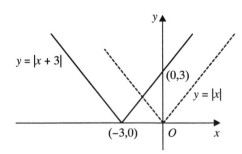

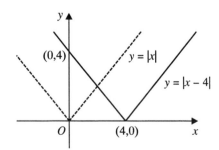

The transformation $y = f(ax)$

Points on the y-axis remain unchanged in this transformation. In general, the point $[t, f(t)]$ maps to the point $[at, f(at)]$, where t is any member of the domain of the function f. The following example shows the effect of taking different values of a.

Example 12

Starting with the basic curve $y = x^3$, sketch the curve $y = (ax)^3$ when $a = 2, \frac{1}{2}, -1, -2$

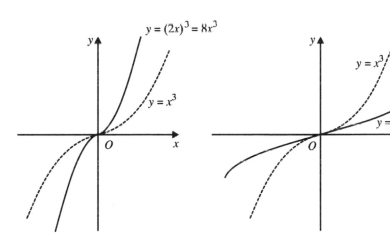

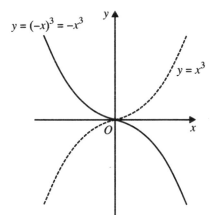
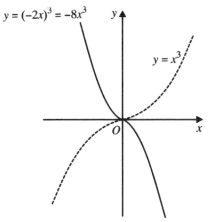

Exercise 3D

In this exercise, take the domain of each function to be the set of real numbers unless informed otherwise.

1 Starting with the graph of the linear function $f(x) = x$, sketch the functions $f(x + 1)$, $f(3x)$, $f(x)$, $f(x) + 1$, $4f(x)$.

2 On the same diagram, for $0 \leqslant x \leqslant 2$, sketch the graphs of the curves $y = x^2$, $y = x^3$, $y = x^4$.

3 Consider the three curves $y = f(x)$, where $f(x) = x^2$, x^3, x^4. For which curve(s) is it true to say that (a) $f(x) = f(-x)$ (b) $f(x) = -f(-x)$ for all values of x?

4 Sketch the graphs of $y = |3x - 2|$ and $y = |x|$. Determine the coordinates of the points of intersection of the two graphs.

5 Given that $f(x) = \dfrac{1}{x}$, $x > 0$, sketch on the same axes the graphs of $y = f(x)$, $y = f(3x)$, $y = 3f(x)$, $y = f(x + 3)$.

6 Given that $f(x) = (x + 2)^2$, sketch on the same axes the graph of $y = f(x)$ with the graph of (a) $y = f(x) - 2$ (b) $y = f(x - 2)$ (c) $y = -2f(x)$ (d) $y = f(2x)$.

7 The functions g and h are defined for all real x by

$$g : x \mapsto 2 + \frac{x}{3}$$
$$h : x \mapsto 3(x - 2)$$

(a) Show that $g^{-1} = h$.
(b) Sketch the graphs of $y = gg(x)$ and $y = hh(x)$.
(c) Sketch the graphs of $y = g(x + 2)$ and $y = h(x - 6)$.

8 Given that $f(x) = x^2$, sketch in separate diagrams the graphs of (a) $y = -f(x)$ (b) $y = f(2 - x)$ (c) $y = 2-f(x)$ (d) $y = f(2x)$ (e) $y = 2f(x)$.

SUMMARY OF KEY POINTS

1. A function is a mapping between two variables, usually called x, the independent variable, and y, the dependent variable. The function, or mapping, is given by f and written as $f : x \mapsto y$. The set of values taken by x, the independent variable, is called the domain and the resulting set of values arising for y is called the range.

2. When defining a function, it is necessary to state both the rule for getting from each x to the corresponding y and the set (domain) of values assigned for x.

3. $f(x)$ is the image of x under the function f. The graph of the function f has equation $y = f(x)$; this is often called the cartesian equation of the function.

4. In forming composite functions $gf(x)$, apply f first, followed by g. Remember that, in general, $fg(x) \neq gf(x)$.

5. For a one–one function f, the inverse function f^{-1} exists, and $ff^{-1}(x) = f^{-1}f(x) = x$.

6. The line $y = x$ is an axis of symmetry for the curves $y = f(x)$ and $y = f^{-1}(x)$.

7. $|x| = x$ for $x \geqslant 0$ and $|x| = -x$ for $x < 0$.

8. For an even function f, $f(x) = f(-x)$.
   For an odd function f, $f(-x) = -f(x)$.

9. Transformations:
   $y = af(x)$: points on x-axis remain unchanged.
   $[t, f(t))] \rightarrow [t, af(t)]$
   $y = f(x) + a$: translates in y^+ direction for $a > 0$,
   in y^- direction for $a < 0$.
   $y = f(x + a)$: translates in x^- direction for $a > 0$,
   in x^+ direction for $a < 0$.
   $y = f(ax)$: all points on y-axis remain unchanged.
   $[t, f(t)] \rightarrow [at, f(at)]$

Review exercise

1 Factorise completely

    (a) $x^2 - 7x$,

    (b) $x^2 - 7x - 18$,

    (c) $(x - 3)^2 - y^2$ [L]

2 (a) Solve the equation $(2x + 3)^2 = 4$.

    (b) Solve the equation $2x^2 + 7x = 11$, giving your answers
         to one decimal place.   [L]

3 Find the product of

$$(x - 1), \ (2x + 3) \text{ and } (x + 4)$$

express your answer in descending powers of x.   [L]

4 Solve the simultaneous equations

$$y - 2x = 1$$
$$y^2 = 2x^2 + x$$

                                              [L]

5 The function g is defined by

$$g : x \mapsto \frac{1}{1 + x}, \text{ where } x \in \mathbb{R}, \ x \neq -1$$

Given that $g(x) = g[g(x)]$, show that $x^2 + x - 1 = 0$. Hence
find, to 2 decimal places, the values of x for which
$g(x) = g[g(x)]$.   [L]

6 Solve the simultaneous equations

$$5a - 2b = 9$$
$$2a - 5b = 12$$

                                            [L]

7 Simplify as far as possible

$$2(3x - 4) - 5(x - 3)$$

                                            [L]

8 Expand and simplify $(2 + \sqrt{5})(2 - \sqrt{5})$.

9 Solve the equation

$$3(x-2) - 5(1-2x) = 15 \qquad \text{[L]}$$

10 (a) Solve the equation

$$2x^2 + x - 2 = 0$$

giving your answers correct to 2 decimal places.

(b) The quadratic equation $x^2 - x + r = 0$ is satisfied when $x = 2$. Find the value of r. [L]

11 List the set of integers n for which

$$2n^2 - 13n + 15 < 0 \qquad \text{[L]}$$

12 (a) Solve the equation

$$5(3x-1) - 4(x+3) = 9x + 5$$

(b) Add together $(-2p - 3q + 5)$ and $(3p - 4q - 6)$, simplifying your answer. [L]

13 The functions f, g are defined by

$$f : x \mapsto 6x - 1, \, x \in \mathbb{R}$$

$$g : x \mapsto \frac{4}{x-1}, x \in \mathbb{R}, \, x \neq 1$$

Find in its simplest form

(a) the inverse function f^{-1}

(b) the composite function fg.

(c) Determine the values of x which satisfy $f(x) = g(x)$. [L]

14 Given that $z = 1$, solve the simultaneous equations

$$x + y + 2z = 7$$
$$2x + 3y + 3z = 16 \qquad \text{[L]}$$

15 Solve the simultaneous equations

$$x - 2y = 1$$
$$x^2 - 2xy + 2y^2 = 25 \qquad \text{[L]}$$

16 Rationalise the denominator of $\dfrac{3 - \sqrt{7}}{\sqrt{7} + 1}$.

17 (a) Solve the equation

$$(2x - 5)^2 = 9$$

(b) Solve the equation $x^2 - 5x - 2 = 0$, giving your answers correct to 2 decimal places. [L]

18 Factorise completely

   (a) $5x^2 - 20x$

   (b) $a^2 - 2a - 24$

   (c) $9x^2 - 1$ [L]

19 Find the set of values of x for which

$$2x^2 + x - 1 < 0$$ [L]

20 Simplify as far as possible

$$(4x + 3y - 5z) - (2x - 3y + 4z)$$ [L]

21 Find the numbers m and n such that

$$5 + 4x - x^2 \equiv m - (x - n)^2$$

for all real values of x. State the range of the function f, where

$$f : x \mapsto 5 + 4x - x^2,\ x \in \mathbb{R}.$$ [L]

22 (a) Factorise completely

   (i) $4x^2 - y^2$

   (ii) $2x^2 - 9x + 10$

   (iii) $2x^3 + 6x$

  (b) Solve the simultaneous equations

$$4a + 5b = 1$$
$$3a - 4b = 24$$ [L]

23 Solve the simultaneous equations

$$2x + 3y = 5$$
$$x^2 + 2xy = 10 + y$$ [L]

24 Rationalise the denominator of $\dfrac{\sqrt{3} - 1}{\sqrt{3} - 3}$.

25 (a) Factorise completely

   (i) $x^2 - 2x - 35$,

   (ii) $4x^2 - 16x + 16$.

  (b) Factorise $a^2 - b^2$ and hence, or otherwise, evaluate

$$5248^2 - 4752^2$$ [L]

26 (a) Solve the equation $x^2 + 4x = 0$.

  (b) Solve the equation $x^2 + 4x = 12$.

  (c) Solve the equation $x^2 + 5x + 1 = 0$, giving your answers correct to 2 decimal places.   [L]

27 Find the set of values of x for which
$$(x - 1)(x - 2) < 6$$ [L]

28 (a) Solve the equation $x^2 + x - 9 = 0$, giving your answers correct to 2 decimal places.

  (b) Multiply $(x^2 + 2x + 1)$ by $(x - 2)$, arranging your answer in descending powers of x.   [L]

29 A function f is defined by
$$\text{f} : x \mapsto 1 - \frac{1}{x}, \quad x \in \mathbb{R}, \, x \neq 0, \, x \neq 1.$$

  Find (a) ff(x)

    (b) fff(x)

    (c) f$^{-1}(x)$.   [L]

30 Solve the simultaneous equations
$$3a + 7b = 28$$
$$a + b = 5\tfrac{1}{3}$$ [L]

31 Given that
$$x + y = 4$$
$$\text{and } x^2 - y = 8$$

  calculate the two possible values of x.   [L]

32 Expand and simplify $(\sqrt{7} - 4)(3\sqrt{7} - 5)$.

33 (a) Factorise completely $2p^2 - 6q + 6p - 2pq$.

  (b) Solve the equation $3x^2 + 3x = 5$, giving your answers to two decimal places.   [L]

34 Solve the equation
$$\frac{3x - 8}{2} = \frac{6x - 4}{5} + 1$$ [L]

35 Find the set of values of x for which
$$(2x + 1)^2 < 9(4 - x)$$ [L]

36 (a) Expand $2(3x - x^2) + 2(x^3 - 3x + 4)$ giving your answer in ascending powers of x.

   (b) Solve the equation $3(2x - 3) - 2(1 - 2x) = 29$. [L]

37 (a) Multiply $(2 - 3x - x^2)$ by $(1 - x)$, arranging your answer in ascending powers of x.

   (b) Solve the equation $x^2 + 16x - 17 = 0$. [L]

38 The functions f and g are defined for $x \in \mathbb{R}$ by

$$f : x \mapsto 2x^2 - 1,$$
$$g : x \mapsto x + 3.$$

   (a) State the range of f.

   (b) Define the composite function fg in the form $fg : x \mapsto$.

[L]

39 Solve the simultaneous equations

$$4x + 7y = 6$$
$$3x - 2y = 19$$

[L]

40 Solve the simultaneous equations

$$x + 2y = 4$$
$$x^2 + xy - 2y^2 = 4$$

[L]

41 Rationalise the denominator of $\dfrac{3\sqrt{11}}{2\sqrt{11} + 5}$.

42 Factorise completely

   (a) $16p^2 - q^2$

   (b) $x^2 - 5x - 24$

   (c) $3ab + 9a - 2b - 6$ [L]

43 Solve the equation

$$x^2 + 7x + 11 = 0$$

giving your answers to 3 significant figures. [L]

44 Find the set of values of x for which

$$2x(x - 2) < x + 3$$ [L]

45 (a) Solve the equation

$$5x^2 - 7x = 3$$

giving your answers to 2 significant figures.

(b)   Multiply $(3x^2 - 2x - 4)$ by $(3 - 5x)$ arranging your answer in ascending powers of x. [L]

46   The functions f and g are defined, for $x \in \mathbb{R}$, by

$$f : x \mapsto \frac{1}{x + 2}, \quad x \neq -2$$

$$g : x \mapsto \frac{x - 4}{x + 2}, \quad x \neq -2 \qquad \text{[L]}$$

Define, in a similar manner, the functions

(a)   fg

(b) g^{-1} [L]

47   Given that $x = -\frac{1}{2}$ and $x = 1$ are *both* solutions of the equation $ax^2 + bx - 1 = 0$, solve the equation in order to find the values of a and b. [L]

48   Solve the simultaneous equations

$$3x + 2y = 25$$

and $$xy = 4 \qquad \text{[L]}$$

49   Expand and simplify $(3\sqrt{3} - 2)(\sqrt{3} - 1)$.

50   (a)   Factorise completely

(i) $x^2 - 12x - 13$

(ii) $4p^2 q - 25q^3$

(iii) $2ac - bc - 4ad + 2bd$

(b)   Solve the simultaneous equations

$$2x - 3y = 11$$

$$3x - 2y = 6\tfrac{1}{2} \qquad \text{[L]}$$

51   (a)   Factorise completely $3x^2 - 3$ and hence simplify

$$\frac{x^2 - 5x + 4}{3x^2 - 3}$$

(b)   Solve the equation $3x^2 - 6x - 10 = 0$, giving your answers to 2 decimal places. [L]

52   Find the set of values of x for which

$$16 - 5x \geqslant 1 - 2x \qquad \text{[L]}$$

53 (a)   Simplify as far as possible

$$2(6x - 5z + 3y) - (5y - 2x + z)$$

(b)   Solve the equation

$$5(3x - 1) - 3(1 - 2x) = 76 \qquad \text{[L]}$$

54 The function f is defined by

$$f : x \mapsto \frac{x + 3}{x - 1}, x \in \mathbb{R}, x \neq 1.$$

Find

(a)   the range of f
(b)   ff(x)
(c)   f$^{-1}(x)$ \qquad [L]

55 (a)   Solve the simultaneous equations

$$x + 3y = -8$$
$$4x + y = 1$$

(b)   Find the coefficient of x^2 in the product of

$$(2x^2 - 3x + 1) \text{ and } (x^2 + 2x - 1) \qquad \text{[L]}$$

56 Solve the simultaneous equations

$$x - y = 2$$
$$x^2 + xy - y^2 = 11 \qquad \text{[L]}$$

57 Rationalise the denominator of $\dfrac{1}{\sqrt{17} - \sqrt{13}}$.

58 Factorise completely
(a) $2x^2 - 18y^2$
(b) $3ac - 6ad - 2bc + 4bd$ \qquad [L]

59 Solve the equation $2x^2 - x - 6 = 0$. \qquad [L]

60 Find the set of values of x for which

$$x^2 - x - 12 > 0 \qquad \text{[L]}$$

61 (a)   Solve the equation $3x^2 - 2x - 7 = 0$, giving your
       answers correct to 3 significant figures.
(b)   Given that

$$(x^2 - 2x + 3)(x^2 + kx - 2) = x^4 - 15x^3 + qx^2 + px - 6,$$

find the values of k, p and q. \qquad [L]

62 The function f is defined by

$$f : x \mapsto \frac{3x + 1}{x - 2}, \quad x \in \mathbb{R}, x \neq 2$$

Find, in a similar form, the functions

(a) ff   (b) f^{-1} [L]

63 Solve the simultaneous equations

$$3x = 11 + 2y$$
$$5x - 3y = 18$$ [L]

64 Solve the simultaneous equations

$$2y - 3x - 2$$
$$4y^2 \quad 4xy - 18x^2 = 5$$ [L]

65 Rationalise the denominator of $\dfrac{2 - 3\sqrt{5}}{2\sqrt{5} + 1}$

66 Factorise completely

(a) $3p^2 - 9p$

(b) $(y + 5)^2 - 16$ [L]

67 (a) Solve the equation $2x^2 - 7x - 3 = 0$, giving your answers correct to 2 decimal places.

(b) Solve the equation

$$\frac{(3x - 8)}{2} = \frac{(6x + 1)}{5}$$ [L] .

68 (a) Solve the equation

$$2x^2 - 7x + 4 = 0$$

giving your answers correct to 3 significant figures.

(b) Multiply $(2x^2 - 3x + 1)$ by $(2 - 3x)$ arranging the answer in descending powers of x.   [L]

69 The functions f and g are defined by

$$f : x \mapsto 2x + 1, \quad x \in \mathbb{R}$$
$$g : x \mapsto \frac{1}{x}, \quad x \in \mathbb{R}, x \neq 0$$

(a) Calculate the value of gf(2).

(b) Find g^{-1}.

(c) Calculate the values of x for which fg(x) = x.   [L]

70   Solve the simultaneous equations

$$3x - 4y = 10$$
$$5x + 2y = 8 \qquad \text{[L]}$$

71   Solve the simultaneous equations

$$x^2 + y^2 = 1$$
$$2x + y = 2 \qquad \text{[L]}$$

72   Factorise completely

(a) $16x^2 - 9y^2$

(b) $3ab - 9a - 3 + b$

(c) $3y^2 + y - 10$ $\qquad \text{[L]}$

73   Solve the equation $5x^2 - 3 = 5x$, giving your answers to 3 significant figures. $\qquad \text{[L]}$

74   Find the values of a, b and c such that

$$a(x + b)^2 + c \equiv 9x^2 + 72x + 128$$

Hence, or otherwise, find the set of values of x for which

$$9x^2 + 72x + 128 \geqslant 0 \qquad \text{[L]}$$

75   Solve the following inequalities and illustrate each answer on a number line.

(a) $\frac{1}{2}x + 2 \geqslant x - 5$

(b) $x^2 + 4 < 20$ $\qquad \text{[L]}$

76   (a)   Solve the equation $(3x - 1)(x - 1) = 5$.

(b)   Solve the equation $2x^2 - 4x - 3 = 0$, giving your answers correct to 2 decimal places. $\qquad \text{[L]}$

77   Functions f and g are defined by

$$\text{f}: \quad x \mapsto \frac{4}{x - 2}, \quad x \in \mathbb{R}, x \neq 2$$
$$\text{g}: \quad x \mapsto x + 2, \quad x \in \mathbb{R}$$

(a)   Show that $\text{gf}: x \mapsto \dfrac{2x}{x - 2}, \quad x \in \mathbb{R}, x \neq 2.$

(b)   Find $(\text{gf})^{-1}$ in the form

$$(\text{gf})^{-1} : x \mapsto \; \ldots \ldots \qquad \text{[L]}$$

78 (a) Solve the equation

$$6x + 17 = 5(2x - 3)$$

(b) Given that $x^2 + y^2 + z^2 = 29$, calculate the two possible values of y when $x = 2$ and $z = -3$.

(c) Add together $(3a - b)$, $(4b + 3c)$ and $(a - 2c)$, simplifying your answer as far as possible. [L]

79 Solve

(a) $x^2 - 8x + 12 = 0$

(b) $x^2 - 25 = 0$

(c) $2x^2 - 7x - 5 = 0$, giving your answers correct to 2 decimal places. |L|

80 Solve

(a) $x^2 + 3x = 0$

(b) $12x^2 - 4x - 5 = 0$

(c) $2x^2 + 6x + 3 = 0$, giving your answers to 2 decimal places. [L]

81 Factorise completely

(a) $9x^2 - 25y^2$

(b) $pr - 3ps + 2qr - 6qs$

(c) $3a^2 - a - 2$ [L]

82 Solve the equation

$$4x^2 + 4x - 7 = 0$$

giving your answers in the form $p \pm q\sqrt{2}$, where p and q are real numbers to be found. [L]

83 A variable number p is such that

$$3 - 2p \leqslant 6 + p \leqslant 3(14 - p)$$

Find the set of possible values of p, writing your answers in the form $m \leqslant p \leqslant n$, where m and n are integers. [L]

84 (a) Given that $y = 4x^2 - 3x$, find y when $x = -2$.

(b) Multiply together

$$(x^2 + 2x + 3) \text{ and } (x - 1),$$

giving your answer in descending powers of x.

(c) Solve the equation

$$x^2 - 2x = 0$$ [L]

85 Two functions f and g are defined by

$$f : x \mapsto \frac{25}{3x - 2}, \quad x \in \mathbb{R}, 1 < x \leqslant 9$$
$$g : x \mapsto x^2, \quad x \in \mathbb{R}, 1 < x \leqslant 3$$

Find

(a) the range of f

(b) the inverse function f^{-1}, stating its domain

(c) the composite function fg, stating its domain

(d) the solutions of the equation $fg(x) = \dfrac{2}{x - 1}$. [L]

86 Solve the simultaneous equations

$$x + 3y = 9$$
$$4x - 2y = 1$$ [L]

87 Solve the simultaneous equations

$$x^2 - 2xy + y^2 = 1$$
$$\text{and } 2x - y + 1 = 0$$ [L]

88 Rationalise the denominator of $\dfrac{1}{\sqrt{5}(\sqrt{19} + \sqrt{3})}$.

Coordinate geometry

4

Coordinate geometry is the study of the geometry of straight lines and curves using algebraic methods. When you graph a line or curve in two dimensions (2D) it exists in a plane   a two-dimensional flat surface. If two axes at right angles to each other are laid on the plane you can identify any point in the plane, using coordinates that show how far the point is from each axis. One axis is usually called the x-axis, and other is called the y-axis. The point where the axes meet is called the origin and is labelled O.

To define a point P in the plane using coordinate geometry you state the perpendicular distance of the point from the y-axis and its perpendicular distance from the x-axis. This defines the position of the point uniquely.

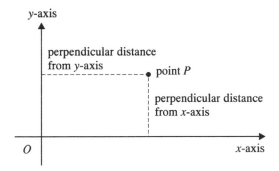

The part of the y-axis that lies above the x-axis is called the positive y-axis and is labelled with positive numbers. The part of the y-axis that lies below the x-axis is called the negative y-axis and is labelled with negative numbers. Similarly, the part of the x-axis to the left of the y-axis is called the negative x-axis and the part to the right is the positive x-axis.

The perpendicular distance of a point in the plane from the y-axis is called the **x-coordinate** (or **abscissa**) of the point. The perpendicular distance of the point from the x-axis is called the **y-coordinate** (or **ordinate**) of the point. These numbers must be labelled with a + or

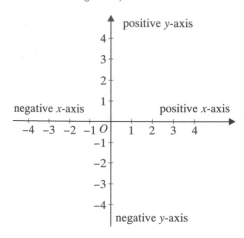

– sign to show where the point lies. The coordinates of a point are written (x, y), where the first number is the x-coordinate.

Example 1

The points A, B, C, D and E are shown with their coordinates on the diagram.

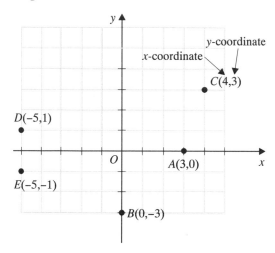

4.1 Finding the distance between two points

Suppose you want to find the distance PQ between the points $P(2,3)$ and $Q(7,5)$. First draw the right-angled triangle PQR. Notice that the x-coordinate of R is 7 as it is the same perpendicular distance from the y-axis as is Q. The y-coordinate of R is 3. That is, R is the point (7,3). The distance PR is $7 - 2 = 5$ units in length. The distance QR is $5 - 3 = 2$ units in length.

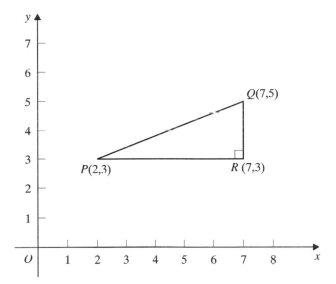

Now that you know the lengths of PR and QR it is easy to calculate the length of PQ using Pythagoras' theorem.

So:
$$PQ^2 = 5^2 + 2^2$$
$$= 25 + 4$$
$$= 29$$

and so $PQ = \sqrt{29} = 5.39$ (3 s.f.).

To find a formula that can always be used to find the distance between two points, you can generalise this process using two points with coordinates (x_1, y_1) and (x_2, y_2).

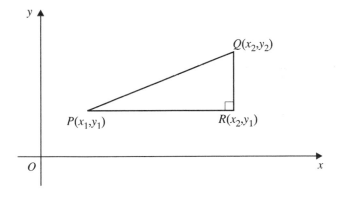

P is the point (x_1, y_1) and Q is the point (x_2, y_2). Drawing the right-angled triangle PQR gives the point R as (x_2, y_1).

The distance PR is $x_2 - x_1$, and the distance QR is $y_2 - y_1$. Using Pythagoras' theorem gives:

$$PQ^2 = (x_2 - x_1)^2 + (y_2 - y_1)^2$$

So:
$$PQ = \sqrt{[(x_2 - x_1)^2 + (y_2 - y_1)^2]}$$

Example 2

Find the distance between the points $P(2,7)$ and $Q(4,10)$.

Using the formula:

$$\begin{aligned}
PQ &= \sqrt{[(x_2 - x_1)^2 + (y_2 - y_1)^2]} \\
&= \sqrt{[(4 - 2)^2 + (10 - 7)^2]} \\
&= \sqrt{(2^2 + 3^2)} \\
&= \sqrt{(4 + 9)} \\
&= \sqrt{13} \\
&= 3.61 \quad \text{(3 s.f.)}
\end{aligned}$$

Example 3

Find the distance between the points $P(-3, -5)$ and $Q(2, -4)$.

Using the formula:

$$\begin{aligned}
PQ &= \sqrt{[(x_2 - x_1)^2 + (y_2 - y_1)^2]} \\
&= \sqrt{[(2 + 3)^2 + (-4 + 5)^2]} \\
&= \sqrt{(5^2 + 1^2)} \\
&= \sqrt{(25 + 1)} \\
&= \sqrt{26} \\
&= 5.10 \quad \text{(3 s.f.)}
\end{aligned}$$

Exercise 4A

Find the lengths of the lines joining:

1 (2,9) and (3,12)
2 (4,5) and (7,10)
3 (2,−3) and (3,8)
4 (−5,7) and (6,9)
5 (2,10) and (−3, −2)
6 (−3, −5) and (−6, −3)
7 (8,7) and (2,3)
8 (9,−4) and (−3, −6)
9 (−2, −6) and (2,8)
10 (−8, −2) and (−3, −4)
11 (−2, 7) and (3,−8)
12 (5,−9) and (2,−7)

13 Show that the triangle ABC is isosceles, where A is the point $(-5, 0)$, B is the point $(-1, 3)$ and C is the point $(2, 7)$.

14 The vertices of a triangle are $A(5, 12)$, $B(-12, 5)$ and $C(-7, 17)$. Show that $\angle ACB$ is $90°$.

15 Calculate the lengths of the sides of the triangle whose vertices are $P(-2, 3)$, $Q(4, 1)$ and $R(-1, -1)$.

16 Two opposite vertices of a square are $P(3, 2)$ and $Q(-5, -10)$. Find the length of:
(a) a diagonal of the square
(b) a side of the square.

17 Show that the point $C(9, 3)$ is at the same distance from $A(2, 2)$ and $B(4, 8)$.

18 Show that the points $(1, -1)$ $(-1, 1)$ and $(\sqrt{3}, \sqrt{3})$ are the vertices of an equilateral triangle.

19 Using the fact that the diagonals of a rectangle are of equal length, show that the points $(6, 0)$, $(2, 4)$, $(3, -3)$ and $(-1, 1)$ are the vertices of a rectangle.

20 The point (a, b) is at the same distance from the origin O, the point $(6, 8)$ and the point $(6, 0)$. Find the values of a and b.

4.2 The gradient of a straight line joining two points

The **gradient** of a straight line is a measure of how steep that line is. The gradient of a line joining two points is defined as:

$$\frac{\text{change in } y\text{-coordinates}}{\text{change in } x\text{-coordinates}}$$

between the two points.

So the gradient joining the points $P(x_1, y_1)$ and $Q(x_2, y_2)$ is:

$$\frac{\text{change in } y\text{-coordinates}}{\text{change in } x\text{-coordinates}} = \frac{y_2 - y_1}{x_2 - x_1}$$

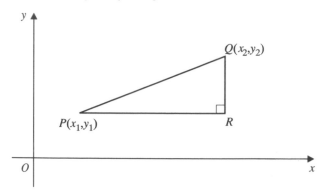

As $y_2 - y_1$ represents the distance QR and $x_2 - x_1$ represents the distance PR, the gradient of the line PQ is actually represented by the ratio

$$\frac{QR}{PR} = \tan \angle QPR$$

(see chapter 7).

Example 4

Find the gradient of the line joining the points (2,5) and (4,8).

Gradient is:

$$\frac{y_2 - y_1}{x_2 - x_1} = \frac{8 - 5}{4 - 2}$$

$$= \frac{3}{2}$$

$$= 1.5$$

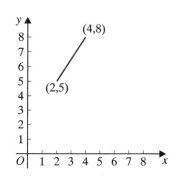

Example 5

Find the gradient of the line joining the points (3,7) and (7,−2).

Gradient is:

$$\frac{y_2 - y_1}{x_2 - x_1} = \frac{-2 - 7}{7 - 3}$$

$$= \frac{-9}{4}$$

$$= -2.25$$

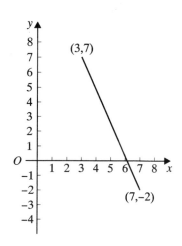

In example 4 the gradient is $+1.5$ and in example 5 it is -2.25. This gives you an idea of the difference in appearance between a line with positive gradient and one with negative gradient. The first line slopes upwards from left to right – it is going 'uphill'. The second line slopes downwards from left to right – it is going 'downhill'. The difference, then, between a straight line with positive gradient and one with negative gradient is that the first slopes 'uphill' and y increases as x increases while the second slopes 'downhill' and y decreases as x increases.

Exercise 4B

Find the gradients of the lines joining the following points:

1 $(4,3)$ and $(8,11)$ 2 $(3,7)$ and $(1,9)$
3 $(6,7)$ and $(3,-2)$ 4 $(5,3)$ and $(10,-5)$
5 $(-8,-7)$ and $(-4,5)$ 6 $(3,4)$ and $(0,13)$
7 $(6,0)$ and $(8,15)$ 8 $(5,-3)$ and $(2,-9)$
9 $(3,9)$ and $(-2,-4)$ 10 $(1,7)$ and $(-6,-14)$
11 $(-2,4)$ and $(7,-3)$ 12 $(-3,-6)$ and $(2,-7)$
13 $(-5,-2)$ and $(-7,-8)$ 14 $(2,-7)$ and $(-7,2)$

15 $A(2,-3)$, $B(7,5)$ and $C(-2,9)$ are the vertices of $\triangle ABC$. Find the gradient of each of the sides of the triangle.

16 $A(3,2)$, $B(6,4)$ and $C(4,0)$ are three points. Find the gradients of the lines AB, BC and CA.

17 $A(2,5)$, $B(4,-2)$ and $C(-7,9)$ are the vertices of $\triangle ABC$. Calculate the gradients of AB, BC and CA.

18 $A(1,2)$, $B(3,1)$ and $C(9,-2)$ are three points. Find the gradient of AB and of AC. What deduction do you make?

19 Three points are $P(-1,-5)$, $Q(1,-2)$ and $R(5,4)$. Find the gradient of PQ and QR. What deduction is possible from your result?

20 Show that the line passing through the points $A(0,-2)$ and $B(3,1\frac{1}{2})$ also passes through the point $C(-6,-9)$.

4.3 The equation of a straight line in the form $y = mx + c$

All straight lines are defined by an equation of the form $y = mx + c$, where m and c are constants. So $y = 2x + 6$, $y = -3x + 2$ and $y = 5x - 3$ each represent a straight line.

For the equation $y = mx + c$, when $x = 0$, $y = c$. Also, when $x = 1$, $y = m + c$.

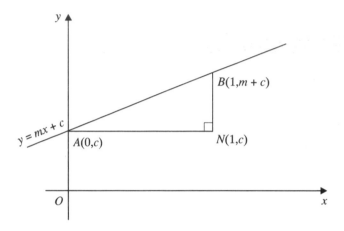

From the diagram you can see that $A(0, c)$ is the point where the line with equation $y = mx + c$ cuts the y-axis. So the distance OA is c. This distance is called the **y-intercept**. Also, since the coordinates of N are $(1, c)$ the gradient of the line AB is

$$\frac{(m + c) - c}{1 - 0} = \frac{m}{1} = m.$$

So m is the gradient of the straight line.

■ **To summarise, if a straight line has equation $y = mx + c$ then: m is the gradient of the line and c is the y-intercept.**

Example 6
Given that the equation of a straight line is $y = -5x + 4$, find

(a) the gradient of the line
(b) the coordinates of the point where the line cuts the y-axis.

Compare $y = -5x + 4$ with $y = mx + c$. Notice that $m = -5$ and $c = 4$. So the gradient of the line is -5 and it cuts the y-axis at $(0,4)$.

Example 7

A straight line has equation $3y + 4x + 6 = 0$. Find

(a) the gradient of the line
(b) the y-intercept.

The equation $3y + 4x + 6 = 0$ can be written as $3y = -4x - 6$ or as $y = -\frac{4}{3}x - 2$.

If you compare this with $y = mx + c$, then $m = -\frac{4}{3}$ and $c = -2$.

So the gradient of the line is $-\frac{4}{3}$ and the y-intercept is -2.

4.4 Finding the equation of a linear graph from the gradient and the y-intercept

You know that if the equation of a straight line (linear) graph is given in the form $y = mx + c$ it has gradient m and y-intercept c. So it is very easy to find the equation of a given straight line if you are told its gradient and its y-intercept. You reverse the process shown in the previous examples to find the equation.

Example 8

A straight line has gradient 3 and cuts the y-axis at the point with coordinates (0,2). Find the equation of the line.

As the gradient is 3 and the y-intercept is 2, the equation of the line is $y = 3x + 2$.

Example 9

A straight line has gradient $-\frac{1}{2}$ and cuts the y-axis at the point with coordinates $(0, -\frac{2}{3})$. Find the equation of the line.

The equation of the line with gradient m and y-intercept c is $y = mx + c$, so the equation of the line with gradient $-\frac{1}{2}$ and y-intercept $-\frac{2}{3}$ is $y = -\frac{1}{2}x - \frac{2}{3}$ or $6y = -3x - 4$ or indeed $6y + 3x + 4 = 0$.

4.5 The equation of a straight line in the form $y - y_1 = m(x - x_1)$

The above is very straightforward. But what happens if you are given the gradient of a straight line and the coordinates of a point on the line instead of its gradient and y-intercept? For example, suppose you are asked to find the equation of the straight line with gradient 3 which passes through the point (2,11). Take A to be the point (2,11) and let $B(x, y)$ be *any* other point on the line. Here is a diagram of the situation:

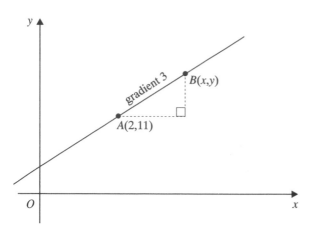

You know that the gradient of the straight line joining the points (x_1, y_1) and (x_2, y_2) is $\dfrac{y_2 - y_1}{x_2 - x_1}$. So the gradient of AB is $\dfrac{y - 11}{x - 2}$. But you are told that the gradient of this line is 3. So:

$$\frac{y - 11}{x - 2} = 3$$

i.e.

$$y - 11 = 3(x - 2)$$
$$y - 11 = 3x - 6$$
$$y = 3x + 5$$

This is the equation of the required straight line.

General method of finding the equation of a straight line from one point and the gradient

Let's try to generalise this method. Suppose you wish to find the equation of the straight line which passes through the point with coordinates (x_1, y_1) and which has a gradient of m. Again, let the point with given coordinates be $A(x_1, y_1)$. Take *any* other point on the line to be B and let it have coordinates (x, y). Here is a diagram of the situation:

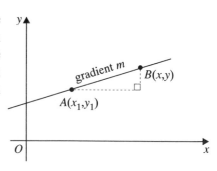

The gradient of AB is:

$$\frac{y - y_1}{x - x_1}$$

As you know that the gradient of the line is m, you have:

$$\frac{y - y_1}{x - x_1} = m$$

or

$$y - y_1 = m(x - x_1)$$

- **This is another important form of the equation of a straight line. Memorise it:**

$$y - y_1 = m(x - x_1)$$

Example 10

Find the equation of the straight line with gradient $-\frac{1}{2}$ which passes through the point (3,2).

Assume that the point (x, y) is any point on the line other than (3,2). Then using the equation

$$\frac{y - y_1}{x - x_1} = m$$

gives:

$$\frac{y - 2}{x - 3} = -\frac{1}{2}$$

or

$$y - 2 = -\frac{1}{2}(x - 3)$$

This can be written: $\qquad 2y - 4 = -x + 3$

or $\qquad 2y + x - 7 = 0$

Example 11

Find the equation of the straight line with gradient $\frac{2}{3}$ which passes through the point $(-1, -4)$.

If the point (x, y) is any point on the line other than $(-1, -4)$ then:

$$\frac{y - y_1}{x - x_1} = m$$

gives:

$$\frac{y - (-4)}{x - (-1)} = \frac{2}{3}$$

$$\frac{y + 4}{x + 1} = \frac{2}{3}$$

or:

$$y + 4 = \frac{2}{3}(x + 1)$$

This can be written:

$$3y + 12 = 2x + 2$$

or:

$$3y - 2x + 10 = 0$$

4.6 The equation of a straight line in the form $\dfrac{y - y_1}{y_2 - y_1} = \dfrac{x - x_1}{x_2 - x_1}$

Finally, let's look at the situation where the gradient of the line is not given: instead, *two* points are given, each of which lies on the line. For example, try to find the equation of line which passes through the points (2,1) and (5,7).

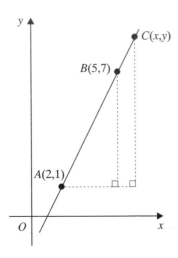

Choose $C(x, y)$ to be any other point on the line *except* (2,1) and (5,7) and use the formula for the gradient of the line joining the two points (x_1, y_1) and (x_2, y_2), which is $\dfrac{y_2 - y_1}{x_2 - x_1}$.

The gradient of the line AC is $\dfrac{y - 1}{x - 2}$ and the gradient of the line AB is $\dfrac{7 - 1}{5 - 2}$.

But ABC is just one straight line and so AC and AB must have the same gradient. Thus:

$$\frac{y - 1}{x - 2} = \frac{7 - 1}{5 - 2} = \frac{6}{3} = 2$$

and:

$$y - 1 = 2(x - 2)$$
$$y - 1 = 2x - 4$$
$$y = 2x - 3$$

General method of finding the equation of a straight line from two points on the line

As before, you can generalise this to find a formula that you can use to solve all such problems. Consider the straight line which passes through the points (x_1, y_1) and (x_2, y_2).

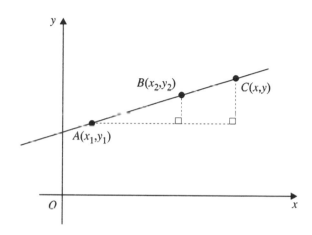

If $C(x, y)$ is any point on the line other than (x_1, y_1) or (x_2, y_2) then the gradient of AC is:

$$\frac{y - y_1}{x - x_1}$$

and the gradient of AB is:

$$\frac{y_2 - y_1}{x_2 - x_1}$$

But as AC is part of the same straight line as AB, it must have the same gradient. So:

$$\frac{y - y_1}{x - x_1} = \frac{y_2 - y_1}{x_2 - x_1}$$

or:

$$\frac{y - y_1}{y_2 - y_1} = \frac{x - x_1}{x_2 - x_1}$$

■ **Memorise this equation.**

Example 12

Find the equation of the line that passes through the points $(2,-1)$ and $(3,7)$.

If (x, y) is any point on the line other than $(2,-1)$ and $(3,7)$ then:

$$\frac{y-(-1)}{7-(-1)} = \frac{x-2}{3-2}$$

$$\frac{y+1}{8} = \frac{x-2}{1}$$

$$y+1 = 8(x-2)$$

or:
$$y+1 = 8x - 16$$

so:
$$y = 8x - 17$$

Example 13

Find the equation of the line which passes through the points $(-4, -2)$ and $(3,6)$.

If (x, y) is any point on the straight line other than $(-4, -2)$ and $(3,6)$ then:

$$\frac{y-(-2)}{6-(-2)} = \frac{x-(-4)}{3-(-4)}$$

$$\frac{y+2}{6+2} = \frac{x+4}{3+4}$$

So:
$$7(y+2) = 8(x+4)$$

i.e.
$$7y + 14 = 8x + 32$$

or
$$7y = 8x + 18$$

You could use a slightly different approach in examples 12 and 13 by finding the gradient of the line and then using the form $y - y_1 = m(x - x_1)$ described on p. 71. The equation found is then checked by using the point (x_2, y_2).

Example 14

Find the equation of the line which passes through the points $(2,-1)$ and $(3,7)$.

$$m = \frac{7-(-1)}{3-2} = 8$$

The equation of the line is

$$y - (-1) = 8(x - 2)$$
$$y + 1 = 8(x - 2)$$

Checking with (3,7) gives:

$$\text{LHS} = 7 + 1 = 8$$
$$\text{RHS} = 8(3 - 2) = 8$$

showing that the equation is correct. You could rearrange the equation into the more usual form:

$$y + 1 = 8(x - 2)$$
$$y + 1 = 8x - 16$$
$$y = 8x - 16 - 1$$

So:
$$y = 8x - 17$$

Example 15

Find the equation of the line which passes through the points $(-4, -2)$ and $(3,6)$.

$$m = \frac{6 - (-2)}{3 - (-4)} = \frac{6 + 2}{3 + 4} = \tfrac{8}{7}$$

The equation of the line is:

$$y - (-2) = \tfrac{8}{7} [x - (-4)]$$
$$y + 2 = \tfrac{8}{7} (x + 4)$$

Checking with (3,6) gives:

$$\text{LHS} = 6 + 2 = 8$$
$$\text{RHS} = \tfrac{8}{7} (3 + 4) = \tfrac{8}{7} \times 7 = 8$$

Once again, you can rearrange the equation into the more usual form:

$$y + 2 = \tfrac{8}{7} (x + 4)$$
$$7(y + 2) = 8(x + 4)$$
$$7y + 14 = 8x + 32$$
$$7y = 8x + 32 - 14$$

So:
$$7y = 8x + 18$$

Exercise 4C

1 Find the gradient of the straight line with equation:

(a) $y = 5x - 7$ (b) $y = -7x + 2$

(c) $y = 4 - 2x$ (d) $y = -7 - x$

(e) $2y = 3x + 4$ (f) $3y = 7x + 1$

(g) $-3y = 4 - 2x$ (h) $-4y = 8 - 3x$

(i) $2y + 3x = 6$ (j) $3y + 4x = 5$

(k) $2y - 5x - 2 = 0$ (l) $5y + 6x - 4 = 0$

2 Find the y-intercept of the straight line with equation:

(a) $y = x - 4$ (b) $y = 2x + 6$

(c) $y = 3 - 5x$ (d) $y = -2 - 4x$

(e) $2y = 3x + 7$ (f) $3y = 7x + 5$

(g) $2y = 6 - 5x$ (h) $5y = 7 - 6x$

(i) $-2y = 3x - 7$ (j) $-5y = 10x - 8$

(k) $3x - 2y - 7 = 0$ (l) $5y - 6x + 20 = 0$

3 Find the coordinates of the point where the following straight lines cut the y-axis:

(a) $y = -3x + 4$ (b) $y = -2x - 7$

(c) $2y = 5x + 3$ (d) $3y = 6x - 2$

(e) $2y + 3x = 5$ (f) $3y - 5x = 9$

(g) $4x + 3y - 7 = 0$ (h) $5x + 4y - 6 = 0$

4 Find the equation of the straight line with gradient 2 and y-intercept 3.

5 Find the equation of the straight line with gradient 5 and y-intercept -4.

6 Find the equation of the straight line with gradient -1 and y-intercept 2.

7 Find the equation of the straight line with gradient $-\frac{2}{3}$ that passes through the point $(0,1)$.

8 Find the equation of the straight line with gradient $-\frac{1}{2}$ that passes through the point $(0,-7)$.

9 Find the equation of the straight line with gradient 2 that passes through $(5,1)$.

10 Find the equation of the straight line with gradient -3 that passes through $(8,3)$.

11 Find the equation of the straight line with gradient $\frac{1}{2}$ that passes through $(-2, 5)$.

12 Find the equation of the straight line with gradient $-\frac{4}{5}$ that passes through $(3, -2)$.

13 Find the equation of the straight line with gradient -7 that passes through the point $(3, 6)$.

14 Find the equation of the straight line with gradient 3 that passes through the point $(1, -\frac{1}{2})$.

15 Find the equation of the straight line through the points

(a) $(2,1)$ and $(5,3)$	(b) $(3,7)$ and $(2,4)$
(c) $(-2,3)$ and $(3,-2)$	(d) $(-1,4)$ and $(2,3)$
(e) $(-1,-7)$ and $(-2,6)$	(f) $(4,-2)$ and $(3,-8)$
(g) $(4,-3)$ and $(6,0)$	(h) $(-1,-5)$ and $(-2,-3)$

SUMMARY OF KEY POINTS

1. The distance between the points (x_1, y_1) and (x_2, y_2) is given by:
$$\sqrt{[(x_2 - x_1)^2 + (y_2 - y_1)^2]}$$

2. The gradient of the line joining (x_1, y_1) and (x_2, y_2) is:
$$\frac{y_2 - y_1}{x_2 - x_1}$$

3. For a straight line with equation $y = mx + c$:
   m represents the gradient of the line
   c represents the y-intercept

4. A straight line with gradient m and y-intercept c has equation $y = mx + c$.

5. A straight line with gradient m which passes through (x_1, y_1) has equation $y - y_1 = m(x - x_1)$.

6. A straight line through the points (x_1, y_1) and (x_2, y_2) has equation:
$$\frac{y - y_1}{y_2 - y_1} = \frac{x - x_1}{x_2 - x_1}$$

Indices

5

This chapter shows you how to manipulate expressions such as a^b. It also introduces exponential functions.

5.1 Index notation

3×3 is usually written 3^2.

$7 \times 7 \times 7 \times 7 \times 7$ is usually written 7^5.

$y \times y \times y \times y \times y \times y \times y \times y$ is usually written y^8.

7^5 in words is: 'seven to the power of five'.

3^6 in words is: 'three to the power of six'.

9^4 in words is: 'nine to the power of four'.

There are two special cases:

a^2 in words is usually read as 'a squared', rather than as 'a to the power of two'

b^3 in words is usually read as 'b cubed', rather than as 'b to the power of three'

In the expression a^b the number a is usually called the **base** and the number b is called either the **power** or the **index** (plural **indices**).

5.2 Multiplying expressions involving indices

To multiply 5^3 by 5^7 you must calculate
$(5 \times 5 \times 5) \times (5 \times 5 \times 5 \times 5 \times 5 \times 5 \times 5) = 5^{10}$.

Similarly, $7^4 \times 7^5 = (7 \times 7 \times 7 \times 7) \times (7 \times 7 \times 7 \times 7 \times 7) = 7^9$.

So $5^3 \times 5^7 = 5^{10} = 5^{3+7}$ and $7^4 \times 7^5 = 7^9 = 7^{4+5}$.

- **In general**

$$x^a \times x^b = x^{a+b}$$

So when multiplying numbers *which have the same base* the indices are added. Remember that, in general, expressions involving indices can only be simplified when the numbers involved have the same base. The expression $5^2 \times 2^3$ can only be simplified by evaluating the expression; in this case, $25 \times 8 = 200$.

Example 1
Simplify $3^9 \times 3^{14}$.

$$3^9 \times 3^{14} = 3^{9+14} = 3^{23}$$

Example 2
Simplify $3x^4 \times 7x^6$.

$$3x^4 \times 7x^6$$
$$= 3 \times x^4 \times 7 \times x^6$$
$$= 3 \times 7 \times x^4 \times x^6$$
$$= 21 \times x^{4+6}$$
$$= 21x^{10}$$

Example 3
Simplify $3x^5 \times 4y^3 \times 2y^2 \times x^3$.

$$3x^5 \times 4y^3 \times 2y^2 \times x^3$$
$$= 3 \times 4 \times 2 \times x^5 \times x^3 \times y^3 \times y^2$$
$$= 24 \times x^{5+3} \times y^{3+2}$$
$$= 24x^8 y^5$$

Example 4
Simplify $2a^2 \times 3a^2 b \times 2b^2 \times 3a^3$

$$2a^2 \times 3a^2 b \times 2b^2 \times 3a^3$$
$$= 2 \times 3 \times 2 \times 3 \times a^2 \times a^2 \times a^3 \times b \times b^2$$
$$= 36 \times a^{2+2+3} \times b^{1+2}$$
$$= 36a^7 b^3$$

Exercise 5A

Simplify:

1 $x \times x \times x \times x$

2 $a \times a \times a \times a \times a \times a$

3 $4 \times a \times a \times a$

4 $6 \times a \times a \times 5 \times b \times b$

5 $x^3 \times x^7$

6 $y^3 \times y^7 \times y^2$

7 $3y^2 \times 4y^5$

8 $2y^2 \times 3y^2 \times 4y^4$

9 $3a^3 \times 4b^2 \times 5b^2 \times a^2$

10 $3x^5 \times y \times 6x^2 \times 2y^3$

11 $6p^3 \times 2p^2 \times 4q^4 \times 2p^5$

12 $6a^3 \times 4b^2 \times 2a^2b^3$

13 $5p^3 \times 2p^2q^3 \times q^4 \times 3q^2$

14 $5x^3 \times 2y^2 \times 3z^3$

15 $2a^3 \times 4ab^2 \times 6bc^3 \times c^4$

16 $8a^3 \times 9b^2$

17 $2a^2 \times 3b^3 \times 6c \times 10a^3b^4c^5$

18 $6p^3 \times 2q^3 \times pq \times 7$

19 $a^3b \times ab^3 \times 2pq \times 3q^2$

20 $\frac{2}{3}ap^3 \times \frac{3}{4}pa^2 \times \frac{1}{2}aq^2 \times \frac{2}{5}qa^2$

5.3 Evaluating $(x^a)^b$

x^3 means $x \times x \times x$.

$(x^3)^2$ means $(x \times x \times x)^2$ which is $(x \times x \times x) \times (x \times x \times x) = x^6$.

Similarly:

$$(a^2)^5 = (a \times a) \times (a \times a) \times (a \times a) \times (a \times a) \times (a \times a) = a^{10}$$

So $(x^3)^2 = x^6 = x^{3 \times 2}$ and $(a^2)^5 = a^{10} = a^{2 \times 5}$.

■ **It is not difficult to see that**

$$(x^a)^b = x^{a \times b} = x^{ab}$$

Example 5

Simplify (a) $(2^2)^3$ (b) $(8^5)^9$ (c) $(x^7)^4$.

(a) $(2^2)^3 = 2^{2 \times 3} = 2^6$

(b) $(8^5)^9 = 8^{5 \times 9} = 8^{45}$

(c) $(x^7)^4 = x^{7 \times 4} = x^{28}$

5.4 Dividing expressions involving indices

The expression $x^4 \div x^2$ means:

$$\frac{\not{x} \times \not{x} \times x \times x}{\not{x} \times \not{x}} = x^2$$

Similarly, $p^9 \div p^5$ means:

$$\frac{\not{p} \times \not{p} \times \not{p} \times \not{p} \times \not{p} \times p \times p \times p \times p}{\not{p} \times \not{p} \times \not{p} \times \not{p} \times \not{p}} = p^4$$

As $x^4 \div x^2 = x^2 = x^{4-2}$ and $p^9 \div p^5 = p^4 = p^{9-5}$ it should be clear that when dividing numbers with the same base, the indices are subtracted.

■ **In general,**

$$x^a \div x^b = x^{a-b}$$

Example 6
Simplify (a) $x^9 \div x^3$ (b) $6x^5 \div 2x^3$ (c) $8a^7 \div 4a^2$ (d) $21x^3y^4 \div 3x^2y$.

(a) $x^9 \div x^3 = x^{9-3} = x^6$

(b) $6x^5 \div 2x^3 = \dfrac{6}{2}x^{5-3} = 3x^2$

(c) $8a^7 \div 4a^2 = \dfrac{8}{4}a^{7-2} = 2a^5$

(d) $21x^3y^4 \div 3x^2y = \dfrac{21}{3}x^{3-2}y^{4-1} = 7xy^3$

Exercise 5B

Simplify:

1 $(a^5)^6$	**2** $(x^2)^7$	**3** $(a^2b^3)^4$
4 $(p^3q^4)^5$	**5** $x^9 \div x^5$	**6** $a^7 \div a^3$
7 $p^{10} \div p^6$	**8** $15p^5 \div 3p^2$	**9** $21a^3 \div 7a$
10 $12a^{15} \div (a^2)^3$	**11** $4(a^3)^5 \div 2a^7$	**12** $16(p^2)^4 \div 4p^3$
13 $8(x^5)^2 : 2x^6$	**14** $20(x^3)^9 : 4(x^2)^{10}$	**15** $15a^2b^7 : 5ab^3$
16 $14a^3b^2c \div 7abc$	**17** $15(a^2b)^3 \div 3ab^2$	**18** $18a^2b^2c^2 \times 2(ab)^2$
19 $3a^2bc^3 \times 4a^2bc \div 2a^2bc^2$	**20** $9xy^4 \times 2x^5yz^3 \div 6(x^2yz)^2$	

5.5 The meaning of x^0

Given that $x^a \div x^b = x^{a-b}$ then $5^4 \div 5^4 = 5^{4-4} = 5^0$.

But $5^4 \div 5^4 = \dfrac{5^4}{5^4} = 1$. So, $5^0 = 1$.

Here is another example: $9^7 \div 9^7 = 9^{7-7} = 9^0$.

But, $9^7 \div 9^7 = \dfrac{9^7}{9^7} = 1$. So, $9^0 = 1$.

In general $$x^n \div x^n = x^{n-n} = x^0$$

But $$x^n \div x^n = \frac{x^n}{x^n} = 1$$

So: $$x^0 = 1$$

5.6 Negative indices

$x^a \div x^b = x^{a-b}$ so $2^3 \div 2^7 = 2^{-4}$.

But: $$2^3 \div 2^7 = \frac{\cancel{2} \times \cancel{2} \times \cancel{2}}{\cancel{2} \times \cancel{2} \times \cancel{2} \times 2 \times 2 \times 2 \times 2} = \frac{1}{2^4}$$

Similarly $$3^5 \div 3^7 = 3^{5-7} = 3^{-2}$$

Now:

$$3^5 \div 3^7 = \frac{\cancel{3} \times \cancel{3} \times \cancel{3} \times \cancel{3} \times \cancel{3}}{\cancel{3} \times \cancel{3} \times \cancel{3} \times \cancel{3} \times \cancel{3} \times 3 \times 3} = \frac{1}{3^2}$$

So $$2^{-4} = \frac{1}{2^4} \quad \text{and} \quad 3^{-2} = \frac{1}{3^2}$$

■ **In general,**

$$x^{-n} = \frac{1}{x^n}$$

5.7 Fractional indices

$x^{\frac{1}{2}} \times x^{\frac{1}{2}} = x^{\frac{1}{2}+\frac{1}{2}} = x^1 = x$.

Also $x^{\frac{1}{3}} \times x^{\frac{1}{3}} \times x^{\frac{1}{3}} = x^{\frac{1}{3}+\frac{1}{3}+\frac{1}{3}} = x^1 = x$.

And $x^{\frac{1}{4}} \times x^{\frac{1}{4}} \times x^{\frac{1}{4}} \times x^{\frac{1}{4}} = x^{\frac{1}{4}+\frac{1}{4}+\frac{1}{4}+\frac{1}{4}} = x^1 = x$ and so on.

It is also true that $\sqrt{x} \times \sqrt{x} = x$ and $\sqrt[3]{x} \times \sqrt[3]{x} \times \sqrt[3]{x} = x$. Also $\sqrt[4]{x} \times \sqrt[4]{x} \times \sqrt[4]{x} \times \sqrt[4]{x} = x$ and so on.

If you compare these two sets of results, you will see that $x^{\frac{1}{2}} = \sqrt{x}$, $x^{\frac{1}{3}} = \sqrt[3]{x}$, and $x^{\frac{1}{4}} = \sqrt[4]{x}$, and so on.

- **In general:**

$$x^{\frac{1}{n}} = \sqrt[n]{x}$$

Notice also that $x^{\frac{2}{3}} = (x^2)^{\frac{1}{3}}$. This is the cube root of x^2 written as $\sqrt[3]{x^2}$.

Similarly $x^{\frac{7}{5}} = (x^7)^{\frac{1}{5}}$. This is the fifth root of x^7 written as $\sqrt[5]{x^7}$.

- **In general:** $\qquad x^{\frac{m}{n}} = (x^m)^{\frac{1}{n}} = \sqrt[n]{x^m}$

  or $\qquad\qquad x^{\frac{m}{n}} = (x^{\frac{1}{n}})^m = (\sqrt[n]{x})^m$

Example 7

Evaluate (a) $9^{\frac{1}{2}}$ (b) $125^{\frac{1}{3}}$ (c) $81^{\frac{3}{2}}$ (d) $36^{-\frac{3}{2}}$.

(a) $9^{\frac{1}{2}} = \sqrt{9} = 3$

(b) $125^{\frac{1}{3}} = \sqrt[3]{125} = 5$

(c) $81^{\frac{3}{2}} = (81^{\frac{1}{2}})^3 = (\sqrt{81})^3 = 9^3 = 729$

(d) $36^{-\frac{3}{2}} = \dfrac{1}{36^{\frac{3}{2}}} = \dfrac{1}{(\sqrt{36})^3} = \dfrac{1}{6^3} = \dfrac{1}{216}$

Exercise 5C

Evaluate:

1 $4^{\frac{1}{2}}$ 2 $16^{\frac{1}{2}}$ 3 $8^{\frac{1}{3}}$ 4 $243^{\frac{1}{5}}$

5 $243^{\frac{3}{5}}$ 6 $8^{-\frac{2}{3}}$ 7 10^0 8 10^{-2}

9 $(-7)^{-3}$ 10 $(6\frac{1}{4})^{\frac{1}{2}}$ 11 $(1\frac{7}{9})^{\frac{1}{2}}$ 12 $(\frac{4}{9})^{-\frac{1}{2}}$

13 $(\frac{8}{27})^{\frac{2}{3}}$ 14 $(-27)^{\frac{2}{3}}$ 15 $(\frac{5}{6})^0$ 16 $(\frac{5}{6})^{-1}$

17 $(\frac{5}{6})^{-2}$ 18 $(15\frac{5}{8})^{\frac{2}{3}}$ 19 $(\frac{16}{25})^{-\frac{1}{2}}$ 20 $(2\frac{7}{81})^{\frac{1}{2}}$

5.8 Exponential functions

A function of the form f: $x \mapsto a^x$ where x is real and a is a positive constant is called an **exponential function**. The word 'exponential' comes from the word **exponent**, which is another word for 'power' or 'index'. So for an exponential function f: $x \mapsto a^x$, $x \in \mathbb{R}$, the variable x is called the power, or the index, or the exponent.

For $a = 2, 3, 4$ here are the graphs of the corresponding exponential functions f: $x \mapsto a^x$:

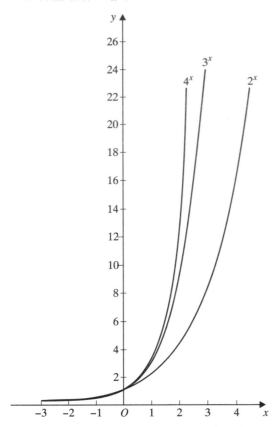

Notice that each graph passes through the point $(0,1)$. It is also true that, as the value of a increases, so the gradient of the corresponding exponential graph increases: the graph of 4^x is much steeper at $x = 2$ than the graph of 2^x at $x = 2$. It can be shown that the gradient of the curve $y = 2^x$ at $(0,1)$ is about 0.693 (see chapter 8), the gradient of the curve $y = 3^x$ at $(0,1)$ is about 1.099, and that of the curve $y = 4^x$ at $(0,1)$ is about 1.386. (This is done on p. 173.)

As the value of a increases so the value of the gradient at $(0,1)$ increases. If the gradient at $(0,1)$ for the curve $y = 2^x$ is about 0.693 and the gradient for the curve $y = 3^x$ at the same point is about 1.099, there must be a value of a, lying between 2 and 3 at which the gradient of the exponential curve at $(0,1)$ is exactly 1. It can be shown that this number is an irrational number that is approximately $2.718\,28\ldots$. The number is symbolised by e. So the gradient of the curve $y = e^x$ at the point $(0,1)$ is exactly 1.

The function f: $x \mapsto e^x, x \in \mathbb{R}$ is called *the* exponential function (as opposed to *an* exponential function).

Exercise 5D

1 Draw the graph of $y = 2^x$. From your graph find to 2 significant figures the value of:
   (a) $2^{1.7}$ (b) $2^{0.3}$ (c) $2^{3.3}$ (d) $2^{-0.8}$

2 Draw the graph of $y = 3^x$. From your graph find to 2 significant figures the value of:
   (a) $3^{2.9}$ (b) $3^{-0.4}$ (c) $3^{1.3}$

3 Draw the graph of $y = e^x$. From your graph, find to 2 significant figures the value of:
   (a) $e^{-0.3}$ (b) $e^{1.9}$ (c) $e^{2.7}$
   Find also the value of x for which: (d) $e^x = 6.7$ (e) $e^x = 16.4$.

5.9 The natural logarithmic function

Here is the graph of the function $f: x \mapsto e^x$, $x \in \mathbb{R}$. You can see that the function is one–one. So it must have an inverse function.

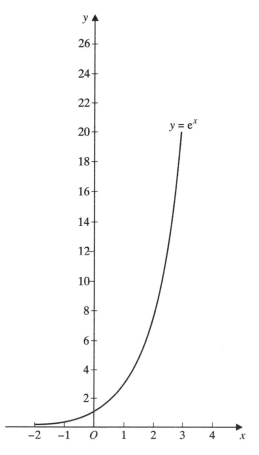

To find the inverse function write $x = e^y$ and try to make y the subject of the equation.

Earlier in this chapter, you learned that in the expression e^y, e is called the base and y is called the power, or index or exponent. A fourth name for y is the **logarithm**. So the expression $x = e^y$ can be interpreted as meaning that y is the logarithm of the number x in the base e. We write this as $y = \log_e x$ or more usually as $y = \ln x$.

Logarithms in base e are frequently called **natural logarithms**.

■ **The function g: $x \mapsto \ln x$, $x > 0$ is the inverse of the function f: $x \mapsto e^x$, $x \in \mathbb{R}$. That is, g $=$ f^{-1}.**

These are the graphs of e^x and of $\ln x$:

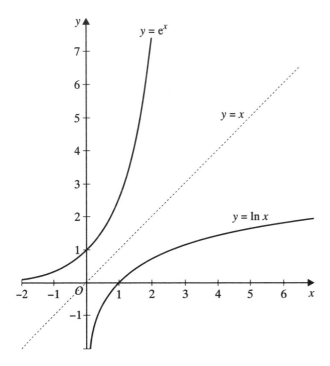

You can see that the graph of $y = \ln x$ is the reflection of the graph of $y = e^x$ in the line $y = x$.

Exercise 5E

1 Draw the graph of $y = \ln x$. Find from your graph the values of
   (a) $\ln 0.4$ (b) $\ln 0.8$ (c) $\ln 1.7$ (d) $\ln 3.3$
   (e) $\ln 4.7$ (f) e^{-1} (g) $e^{-0.6}$ (h) $e^{1.8}$

2   Draw on the same axes the graphs of:
    (a) $y = \ln x$ (b) $y = \ln 3x$ (c) $y = \ln 7x$
    What do you notice?

SUMMARY OF KEY POINTS

1   $a^m \times a^n = a^{m+n}$

2   $(a^m)^n = a^{mn}$

3   $a^m \div a^n = a^{m-n}$

4   $a^0 - 1$

5   $a^{-m} = \dfrac{1}{a^m}$

6   $a^{\frac{1}{m}} = \sqrt[m]{a}$

7   $a^{\frac{m}{n}} = \sqrt[n]{a^m}$

8   f: $x \mapsto e^x$, $x \in \mathbb{R}$, is called the exponential function.

9   The gradient of the curve $y = e^x$ at $(0,1)$ equals 1.

10 The inverse of the function f: $x \mapsto e^x$, $x \in \mathbb{R}$ is
    $f^{-1}: x \mapsto \ln x$, $x \in \mathbb{R}$, $x > 0$.

Sequences and series

6

6.1 Sequences and general terms

A sequence is a succession of numbers formed by following a rule. The numbers in a sequence are sometimes called **terms**. For example, the sequence of numbers produced by the rule $2n + 2$ is:

$$4, \quad 6, \quad 8, \quad 10, \quad 12 \ldots$$

first term   second term

The nth term is also called the **general term** of the sequence because it can be used to find any term in the sequence. When the general term of a sequence is given in terms of n you can find the first, second, third and fourth terms of the sequence by substituting $n = 1$, $n = 2$, $n = 3$ and $n = 4$ in the general term.

Example 1

The nth term of a sequence is $2n - 3$. Find the first four terms and the 15th term.

Take $n = 1, 2, 3$ and 4 in turn to find the terms

$$-1, \quad 1, \quad 3, \quad 5,$$

which are the 1st, 2nd, 3rd and 4th terms of the sequence.

Taking $n = 15$ gives $30 - 3 = 27$, which is the 15th term of the sequence.

Example 2

The first four terms of a sequence are 3, 12, 27, 48. Find a possible nth term.

Rewrite the terms as:

$$3 \times 1, \quad 3 \times 4, \quad 3 \times 9, \quad 3 \times 16$$

That is:
$$3 \times 1^2, \quad 3 \times 2^2, \quad 3 \times 3^2, \quad 3 \times 4^2$$

From this pattern, you can see that $3n^2$ is a possible nth term for the sequence with first four terms 3, 12, 27, 48.

Notice that the nth terms of the sequences discussed in examples 1 and 2 get progressively larger as n gets larger. Such sequences are called **divergent**.

Example 3

Discuss the sequences whose nth terms are:

(a) $(-1)^n$ (b) $(\frac{1}{2})^n$ (c) $\cos(60n°)$.

(a) The terms of the sequence are -1, 1, -1, 1, ... The terms **oscillate** between -1 and 1 successively.

(b) The terms of the sequence are $\frac{1}{2}$, $\frac{1}{4}$, $\frac{1}{8}$, $\frac{1}{16}$, $\frac{1}{32}$, ... Each term is half of the immediately preceding term. As n increases, succeeding terms get progressively nearer to zero.

Sequences whose nth term approaches a finite number as n approaches infinity are called **convergent** – they converge on or get closer to a number. The number they converge on is sometimes called the **limit** or the limiting value.

(c) If you use your calculator you will find the first six terms of the sequence are $\frac{1}{2}$, $-\frac{1}{2}$, -1, $-\frac{1}{2}$, $\frac{1}{2}$, 1, $\frac{1}{2}$, $-\frac{1}{2}$, -1, ... and these terms then repeat from the 7th term to the 12th term, from the 13th term to the 18th term and so on. A sequence that repeats in a set number of terms is called a **periodic sequence**. Here period means the number of terms before the sequence repeats. This sequence is periodic over six successive terms.

Representing a sequence graphically

It is often useful to represent the first few terms of a sequence such as u_1, u_2, u_3, u_4, ... graphically. You can do this by plotting points whose coordinates are $(1, u_1)$, $(2, u_2)$, $(3, u_3)$, $(4, u_4)$, ... These graphs show the first few terms of the three sequences discussed in example 3. The graphs can help you understand the behaviour of a sequence.

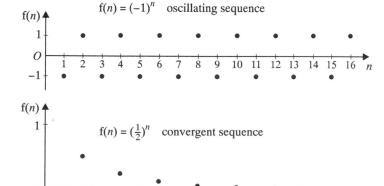

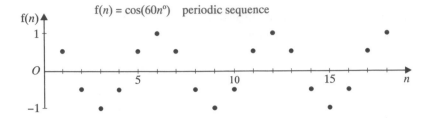

$f(n) = \cos(60n°)$ periodic sequence

Exercise 6A

1   Write down the first four terms of the sequences whose nth terms are

(a) $3n$ (b) $5n - 4$ (c) $2n^2 - 3$ (d) 2^{n-1}

2   Write down possible 5th and 6th terms for the sequence whose first four terms are given. Find a possible nth term.

(a) 3, 5, 7, 9, . . .

(b) 3, 9, 27, 81, . .

(c) $\frac{1}{2}, \frac{2}{3}, \frac{3}{4}, \frac{4}{5}, \ldots$

(d) 5, 9, 17, 33, . . .

(e) 2, 6, 12, 20, . . .

3   The 1st, 2nd and 3rd terms of a sequence are 1, 2 and 5. When asked to give the 4th term, Anne says it is 10, Brian says it is 14 and Cleo says it is 26. Find the 5th and 6th terms for each of the three students' sequences.

4   The nth term of a sequence is n^{-2}. Write down the first four terms. Is this sequence convergent? Give a reason for your answer.

5   Using your calculator, discuss the sequences whose nth terms are

(a) $\cos(180n° - 60°)$ (b) $\sin(45n°)$ (c) $(\sin 45°)^n$.

6   Find the first four terms of the sequence whose nth term is

$$\frac{n-1}{n+1}$$

Investigate what happens to the nth term of this sequence as n increases.

7 Investigate the following sequences to determine whether or not they converge. State the limiting value of those sequences which converge.

(a) $4, 3\frac{1}{2}, 3\frac{1}{3}, \ldots, 3 + \frac{1}{n}, \ldots$

(b) $3, 2, 1, \ldots, 4 - n, \ldots$

(c) $3, 2, \frac{4}{3}, \ldots, 3(\frac{2}{3})^{n-1}, \ldots$

(d) $3, -2, \frac{4}{3}, \ldots, 3(-\frac{2}{3})^{n-1}$

(e) $3, -1, 3, \ldots, 1 + 2(-1)^{n+1}, \ldots$

8 In the following sequences the terms successively increase, or decrease, by a fixed value. Determine the missing terms and find a general term.

(a) $8, \underline{\quad}, \underline{\quad}, 29, \ldots$

(b) $-1, -7, \underline{\quad}, \underline{\quad}, \ldots$

(c) $5, \underline{\quad}, -11, \underline{\quad}, \ldots$

9 At the start of its motion, a pendulum swings through an angle of 30°. Each successive angle of swing is 1% less than the size of the angle in the immediate last swing. Starting with 30°, find the first four terms in the angle-of-swing sequence.

Find also the 20th and the 100th angles of swing.

6.2 Sequences and recurrence relations

You can define the sequence $u_1, u_2, u_3, \ldots, u_n \ldots$ by using a rule or formula for the nth term, u_n, in terms of n, as shown in section 6.1 (p. 90).

A sequence can also be defined by one or more of the first few terms, together with a general relationship between two or more terms of the sequence. This type of relationship is called a **recurrence relation**.

Example 4

A sequence is defined by the recurrence relation:

$$u_n = u_{n-1} + u_{n-2}, \ n > 2$$

where $u_1 = 1$ and $u_2 = 1$.

Determine the 3rd, 4th and 10th terms of the sequence.

For $n = 3$,

$$u_3 = u_2 + u_1 = 1 + 1 = 2$$

For $n = 4$, $\qquad\qquad u_4 = u_3 + u_2 = 2 + 1 = 3$

Continuing this process:

$$u_5 = 5,\ u_6 = 8,\ u_7 = 13,\ u_8 = 21,\ u_9 = 34$$

which gives the 10th term, $u_{10} = 55$.

This sequence in which each term is the sum of the two preceding terms is very well known. It is called the **Fibonacci sequence** after the Italian mathematician who investigated it. The nth term u_n is very complicated and you do not need to know how to find it for the P1 course.

Example 5

Investigate the sequences obtained for the following recurrence relation when (a) $u_1 = 5$ (b) $u_1 = 1$:

$$u_n = \tfrac{1}{2}u_{n-1} + 1,\ n > 1$$

(a) Starting with $u_1 = 5$ and using the recurrence relation gives:
$u_2 = 3.5$, $u_3 = 2.75$, $u_4 = 2.375$, $u_5 = 2.1875$ and $u_6 = 2.09375$.

These terms form a decreasing sequence of numbers that appear to be getting successively nearer to 2. Continuing the process $u_{10} = 2.0058594$, which does not prove the hypothesis but confirms that it is along the right lines.

Taking $u_1 = 2 + s$, where s is a positive number, gives:

$$u_2 = 2 + \frac{s}{2},\ u_3 = 2 + \frac{s}{4} \ldots \text{and } u_n = 2 + \frac{s}{2^{n-1}}$$

The terms $\dfrac{s}{2}, \dfrac{s}{4}, \dfrac{s}{8}, \ldots$ are halving with each step and it is clear that this is a decreasing sequence in which the value of the nth term approaches the value 2 as n approaches infinity.

(b) Starting with $u_1 = 1$ and using the recurrence relation gives $u_2 = 1.5$, $u_3 = 1.75$, $u_4 = 1.875$, $u_5 = 1.9375$, $u_6 = 1.96875$.

This is an increasing sequence of numbers which are getting successively nearer to 2. This can be proved by a method similar to that given in (a).

Exercise 6B

In questions 1–5, find the first four terms of the sequence and then investigate further to decide whether or not the sequence is convergent. For any convergent series, find the limiting value.

1 $u_n = 2u_{n-1} - 4$, $u_1 = 3$, $n > 1$

2   $u_n = 2u_{n-1} + 4$, $u_1 = 3$, $n \geqslant 1$

3   $u_n = \frac{1}{2}(u_{n-1} + 2)$, $u_1 = 0$, $n \geqslant 1$

4   $u_n = \frac{1}{3}u_{n-1} + 1$, $u_1 = 3$, $n \geqslant 1$

5   $u_n = \frac{1}{5}(u_{n-1} - 1)$, $u_1 = 2$, $n \geqslant 1$

6   A sequence is given by the recurrence relation:

$$u_n = 1 + \frac{1}{u_{n-1} + 1}, \quad u_1 = 1, \ n \geqslant 1$$

Work out the 2nd, 3rd and 4th terms of the sequence and find the limiting value of the sequence.

7   In a sequence of increasing numbers, the differences between consecutive terms are 1,2,1,2,1 . . . The 1st, 2nd and 3rd terms are 5, 6, and 8.

Find the 20th, 40th and 100th terms of the sequence. Explain the reasoning in your method.

8   For the sequence u_1, u_2, u_3, . . . , u_n, . . . the terms are related by $u_n = u_{n-1} + 2u_{n-2}$ where $n > 1$, $u_1 = 2$ and $u_2 = 5$.

Find the values of u_7, u_{11} and u_{14}.

9   The sequence given by t_1, t_2, t_3, . . . , t_n, . . . has $t_n = kt_{n-1}$, $n > 1$ and k is given below. Given that $t_1 = 1$, investigate the sequence generated when:

(a) $k = 2$ (b) $k = \frac{1}{2}$ (c) $k = (-1)^n$ (d) $k = \left(-\frac{1}{2}\right)^n$

Which if any of these sequences are convergent?

10   The sequence u_1, u_2, u_3, . . . , u_n, . . . is defined by the relation $u_n = 2u_{n-1} + 6$, $n \geqslant 1$. Given that $u_n = u_1$ for all positive integers n, find the value of u_1.

11   A sequence of numbers v_n, where $n = 1, 2, 3, \ldots$ is given by the recurrence relation

$$v_n = \sqrt{(v_{n-1})}, \quad v_1 = 100$$

Show that this sequence converges, and find the limiting value. Discuss also the convergence of this sequence when $v_1 = \frac{1}{100}$ and find the limiting value.

12   The 1st and 2nd terms of a sequence are 1 and 3. Each successive term is the product of the immediately preceding two terms. For example, the 3rd term is 3 and the 4th is 9. Write down a recurrence relation for the sequence and find the 10th and 13th terms.

6.3 Series and the Σ notation

A sequence consists of a set of terms that follow each other in definite order, such as:

$$u_1,\ u_2,\ u_3,\ \ldots,\ u_n \ldots$$

If you write down the *sum* of the first n terms of the sequence you obtain:

$$u_1 + u_2 + u_3 + \ldots + u_n$$

This is called a **finite series** of n terms. This series is often written as

$$\sum_{r=1}^{n} u_r$$

which means 'the sum of the terms obtained by substituting 1, 2, 3, . . . , n in turn for r in u_r'. The sigma sign Σ is a Greek capital letter S to stand for sum. The numbers below and above it show the lower and upper limits between which the variable is being summed.

If the terms continue infinitely without stopping then

$$u_1 + u_2 + u_3 + \ldots + u_n + \ldots$$

is called an **infinite series**.

Example 6
Write down the first four terms and the nth term of the series

$$\sum_{r=1}^{n} (2r^2 - 1).$$

Taking $r = 1, 2, 3$ and 4 in $2r^2 - 1$ gives:

$$1 + 7 + 17 + 31$$

as the first four terms. The nth term of the series is $2n^2 - 1$.

Example 7
The first four terms of a finite series of n terms are

$$2 + 6 + 12 + 20$$

Write down the next two terms and write the rth term as a function of r, f(r), of the series.

Express the series in the form

$$\sum_{r=1}^{n} f(r)$$

Factorising each term shows that the first four terms of the series may be written as

$$(1 \times 2) + (2 \times 3) + (3 \times 4) + (4 \times 5)$$

Continuing this pattern, the next two terms are 5×6 and 6×7: that is, 30 and 42, respectively. Similarly, the rth term is $r(r + 1)$

The series may be written as

$$\sum_{r=1}^{n} r(r + 1)$$

Exercise 6C

Write down the first four terms and the nth term of the series whose rth term is:

1 $3r$ **2** $2r - 1$ **3** r^3 **4** 2^r **5** $\dfrac{r}{2r - 1}$

Write down the first four terms of these series:

6 $\displaystyle\sum_{r=1}^{n} (3r - 2)$ **7** $\displaystyle\sum_{r=1}^{n} (r + 1)(r + 2)$ **8** $\displaystyle\sum_{r=1}^{n} (2^r - r)$

9 $\displaystyle\sum_{r=1}^{n} \left(\frac{1}{r}\right)$ **10** $\displaystyle\sum_{r=4}^{n} (3r + 1)$

Write down all the terms of these series:

11 $\displaystyle\sum_{r=1}^{6} (2r + 1)$ **12** $\displaystyle\sum_{r=3}^{7} r^2$ **13** $\displaystyle\sum_{r=4}^{8} 3(2^{r-1})$

14 $\displaystyle\sum_{r=100}^{106} (3r - 7)$ **15** $\displaystyle\sum_{r=7}^{10} \left(\frac{2r - 1}{2r + 1}\right)$

6.4 Arithmetic series

An **arithmetic series** is a series in which each term is obtained from the previous term by adding to it or taking from it a constant quantity. The constant quantity is called the **common difference**, because the difference between any two consecutive terms (terms that follow one another) is the same.

Example 8

Find the common difference and the nth term of the arithmetic series $3 + 8 + 13 + \ldots$

The common difference is $8 - 3 = 5$

The first three terms of the series can be written as:

$$3 + (3 + 1 \times 5) + (3 + 2 \times 5)$$

Continuing this pattern the nth term is

$$3 + (n - 1) \times 5 = 3 + 5n - 5$$

that is, $5n - 2$. A quick check shows that $5n - 2$ gives $3 + 8 + 13$ when n is taken as 1, 2 and 3 in turn. Generalising gives:

$$a + (a + d) + (a + 2d) + (a + 3d) + \ldots \text{ to } n \text{ terms}$$

This is the standard way of generalising an arithmetic series with first term a and common difference d. By considering the pattern of terms, you can see that the nth term of the series is:

■ $$n^{\text{th}} \text{ term} = a + (n - 1)d$$

The sum of an arithmetic series

Example 9

Find the sum of the first 10 terms of the arithmetic series $6 + 10 + \ldots$

First notice that the common difference is 4 and that the last term (the 10th term in this case) is $6 + 9 \times 4 = 42$. The series is:

$$6 + 10 + 14 + \ldots + 38 + 42$$

Written backwards the series is:

$$42 + 38 + 34 + \ldots + 10 + 6$$

Adding the corresponding terms in these two series gives:

$$48 + 48 + 48 + \ldots + 48 + 48$$

That is, *twice* the sum of the series is $10 \times 48 = 480$. The sum of the series is 240.

The sum of *any* arithmetic series

You can use the method just shown to find the sum S_n of the first n terms of the general arithmetic series

$$S_n = a + (a+d) + (a+2d) + \ldots + (L-d) + L$$

where L is the nth term (L for 'last').

Rewriting the series in the reverse order gives:

$$S_n = L + (L-d) + (L-2d) + \ldots + (a+d) + a$$

Add these two together to get:

$$2S_n = (a+L) + (a+L) + \ldots + (a+L)$$

That is,

$$2S_n = n(a+L) \quad \text{and} \quad S_n = \frac{n}{2}(a+L).$$

Now $L = a + (n-1)d$ because L is the nth term of the series.

Substituting for L in the expression for S_n gives:

■
$$S_n = \frac{n}{2}[2a + (n-1)d]$$

You can use this formula to find the sum of any arithmetic series to n terms if you know the first term and the common difference of the series.

Example 10

Find the sum of the first 1000 natural numbers.

The series $1 + 2 + 3 + \ldots + 1000$ is arithmetic. Using the formula $S_n = \frac{n}{2}(a+L)$, with $a = 1$, $L = 1000$ and $n = 1000$ gives:

$$S_{1000} = 500(1 + 1000) = 500\,500.$$

The sum of the first 1000 natural numbers is $500\,500$.

Example 11

Find the sum of the first 22 terms of the arithmetic series $6 + 4 + 2 + \ldots$

Use the formula $S_n = \frac{n}{2}[2a + (n-1)d]$, with $n = 22$, $a = 6$ and $d = -2$. This gives:

$$\begin{aligned}
S_{22} &= 11[12 + 21(-2)] \\
&= 11 \times [-30] \\
&= -330
\end{aligned}$$

The sum of the first 22 terms is -330.

Example 12

Evaluate $\displaystyle\sum_{r=1}^{16}(5r-1)$.

The first three terms are found by taking $r=1$, 2 and 3, and $r=16$ gives the last term. This gives the arithmetic series $4+9+14+\ldots+79$, which has 16 terms. Use the formula $S_n = \dfrac{n}{2}(a+L)$ with $n=16$, $a=4$ and $L=79$.

$$S_{16} = 8(4+79) = 664$$

Example 13

The 3rd and 7th terms of an arithmetic series are 71 and 55. Find:
(a)   the first term and the common difference
(b)   the sum of the first 45 terms.

(a)   For the arithmetic series with first term a and common difference d,

$$3\text{rd term} = a+2d = 71$$
$$7\text{th term} = a+6d = 55$$

Solving these equations simultaneously gives:

$$4d = -16, \ d = -4$$

$$a = 55 - (-24) = 79$$

The first term is 79 and the common difference is -4.

(b)   Use the formula $S_n = \dfrac{n}{2}[2a+(n-1)d]$ with $n=45$, $a=79$ and $d=-4$.

$$S_{45} = \frac{45}{2}[158 + 44 \times (-4)] = \frac{45}{2}[-18] = -405$$

The sum of the first 45 terms is -405.

Example 14

The sum of the first n terms of a series is $n^2 + 3n$.
(a)   Find the nth term of the series.
(b)   Show that the series is arithmetic and find the common difference.

(a) The nth term of the series $= S_n - S_{n-1}$

$$= n^2 + 3n - (n-1)^2 - 3(n-1)$$
$$= n^2 + 3n - (n^2 - 2n + 1) - 3n + 3$$
$$= n^2 + 3n - n^2 + 2n - 1 - 3n + 3$$
$$= 2n + 2$$

(b) Substituting $n = 1,2,3, \ldots$ in turn in the nth term, $2n + 2$, gives the series:

$$4 + 6 + 8 + 10 + \ldots$$

This is clearly an arithmetic series with first term 4 and common difference 2.

Exercise 6D

In questions 1–5 find the common difference and the sum of the arithmetic series to the number of terms stated.

1 $1 + 4 + 7 + \ldots$ 17 terms

2 $-3 + 1 + 5 + \ldots$ 21 terms

3 $19 + 12 + 5 + \ldots$ 16 terms

4 $0.25 + 0.5 + 0.75 + \ldots$ 30 terms

5 $£50 + £75 + £100 + \ldots$ 66 terms

Evaluate each of the following:

6 $\displaystyle\sum_{r=1}^{23} r$ **7** $\displaystyle\sum_{r=1}^{32} (r - 3)$ **8** $\displaystyle\sum_{r=7}^{13} (3r - 2)$

9 Find the sum of all positive even numbers less than 200.

10 Find the sum of all positive odd numbers between 100 and 400.

11 An arithmetic series has 1st term 10 and 17th term 122. Find the common difference and the sum of the first 20 terms.

12 In an arithmetic series the 20th term is 200. The common difference of this series is -6. Find the first term and the sum of the first 19 terms.

13 In an arithmetic series the 8th term is 3 and the 3rd term is 8. Find the first term, the common difference and the sum of the first 10 terms.

14 An arithmetic series has first term 5 and common difference 3. Starting with the first term, find the least number of terms that have a sum greater than 1500. Starting with the first term, find also the number of terms that have a sum of 3925.

15 The boring of a well costs £5 for the first metre depth, £11 for the second metre and £17 for the third metre. The costs for successive 1 metre depths continue in the arithmetic series which has £5 as the first term and £6 as its common difference. Find the cost of boring a well of depth (a) 50 m (b) 100 m.

16 The sum of the first n terms of a series is $3n^2 - 2n$. Find the 1st, 10th and nth terms of the series.

17 The sum of the first n terms of an arithmetic series is S_n. $S_{20} = 45$ and $S_{40} = 290$. Find the first term and the common difference of the series. Find also the sum of the first 60 terms of the series.

18 The 3rd, 4th and 5th terms of an arithmetic series are $(4 + x)$, $2x$ and $(8 - x)$. Find the value of x, and the sum of the first 24 terms of the series.

19 The 5th and 7th terms of an arithmetic series are $\frac{1}{6}$ and $\frac{1}{2}$. Find the rth term and the sum of the first 18 terms of the series.

20 The first and last terms of an arithmetic series are -12 and 22. The sum of all the terms is 260. Find the number of terms in the series.

21 Given that $\displaystyle\sum_{r=1}^{2n} (4r - 1) = \sum_{r=1}^{n} (3r + 59)$, find the value of n.

22 An arithmetic series has common difference d where $d > 0$. Three consecutive terms of the series, $x - d$, x, and $x + d$, have a sum of 24 and a product of 120. Calculate the value of d.

23 The sums of the first four terms and the last four terms of an arithmetic series are 2 and -18. The 5th term is -2. Find the common difference and the number of terms in the series.

24 Find the sum of all the integers between 0 and 200 that are not divisible by 4.

25 The sum of the 1st and 2nd terms of an arithmetic series is x and the sum of the $(n - 1)^{\text{th}}$ and nth terms is y. Show that the sum of the first n terms is $\frac{n}{4}(x + y)$. Find an expression for the common difference in terms of x, y and n.

6.5 Geometric series

A series in which each term is obtained by multiplying the previous term by a fixed number r, where r can take any value except 0, 1, or -1, is called a **geometric series**.

The number r is called the **common ratio** of the series because the ratio of any term to the term before it in the series is constant.

Example 15

Find the common ratios and the nth terms of these geometric series:

(a) $1 + 3 + 9 + \ldots$
(b) $125 + 25 + 5 + \ldots$
(c) $16 - 8 + 4 - \ldots$

(a) The common ratio is 3 because each term is 3 times the previous term. The series can be written as:

$$1 + 3 + 3^2 + \ldots$$

The nth term is 3^{n-1}.

(b) The common ratio is $\frac{1}{5}$ because each term is one-fifth the previous term. The series can be written as:

$$5^3 + 5^2 + 5 + \ldots$$

The nth term is 5^{4-n}.

(c) The common ratio is $-\frac{1}{2}$ because each term is $-\frac{1}{2}$ times the previous term. The series can be written as:

$$2^4 - 2^3 + 2^2 - \ldots$$

The nth term $= 2^4 \times \left(-\frac{1}{2}\right)^{n-1}$

$$= (-1)^{n-1} \times 2^{5-n}$$

For each nth term that you find, check by putting $n = 1, 2, 3 \ldots$ that the correct series is generated.

Here is the standard way of representing a general geometric series:

$$a + ar + ar^2 + \ldots$$

The first term is a and the common ratio is r. The nth term of this series is ar^{n-1}.

The sum of a geometric series

The sum of a geometric series can be written:

$$S = a + ar + ar^2 + \ldots + ar^{n-1}$$

Muliplying by r gives: $\quad rS = \quad ar + ar^2 + \ldots + ar^{n-1} + ar^n$

Subtracting the expression for rS from the expression for S gives:

$$S - rS = a - ar^n$$

That is: $\quad\quad\quad\quad\quad\quad S(1 - r) = a(1 - r^n)$

Dividing by $1 - r$ gives: $\quad\quad S = \dfrac{a(1 - r^n)}{1 - r}$

This is the standard formula for the sum of the first n terms of a geometric series with first term a and common ratio r. It may also be written as

$$S = \frac{a(r^n - 1)}{r - 1}$$

6.6 Geometric series that converge

The sum to n terms of the geometric series with first term a and common ratio r is:

$$\frac{a(1 - r^n)}{1 - r} = \frac{a}{1 - r} - \frac{ar^n}{1 - r}$$

If r lies between -1 and 0 or between 0 and 1, then r^n gets smaller and smaller as n increases. [For example, $\left(\frac{1}{2}\right)^2 = \frac{1}{4}$, $\left(\frac{1}{2}\right)^3 = \frac{1}{8}$, $\left(\frac{1}{2}\right)^4 = \frac{1}{16} \ldots$] This means that you can make the term $\dfrac{ar^n}{1 - r}$ as small as you like provided you take a large enough value of n. As n approaches infinity so $\dfrac{ar^n}{1 - r}$ approaches zero. In other words the values of $\dfrac{ar^n}{1 - r}$ form a decreasing sequence which has the **limiting value** zero as n approaches infinity.

The two statements $-1 < r < 0$ and $0 < r < 1$, can be combined by writing $|r| < 1$. This reads 'mod r is less than 1'. Mod r, written $|r|$, means the absolute value of r, or the actual numerical value of r. For example, if $r = 3$, $|r| = 3$ but if $r = -3$ then $|r| = 3$ too. (There is more about this in section 3.4.)

So if $|r| < 1$, the geometrical series:

$$a + ar + ar^2 + \ldots + ar^{n-1} + \ldots$$

has the sum to n terms:

$$S_n = \frac{a(1 - r^n)}{1 - r}$$

which converges to the value $\dfrac{a}{1 - r}$ as n approaches infinity. The series has a **sum to infinity** of $\dfrac{a}{1 - r}$, provided that $|r| < 1$.

Example 16

Find to the nearest integer the sum of the first 11 terms of the geometric series with 1st term 5 and 2nd term 6.

$$a = 5, \ ar = 6 \quad \text{and so} \quad r = \frac{ar}{a} = \frac{6}{5}$$

The common ratio of the series is $\frac{6}{5} = 1.2$. With $a = 5$, $r = 1.2$ and $n = 11$, using the formula

$$S_n = \frac{a(1 - r^n)}{1 - r}$$

gives:
$$S_{11} = \frac{5(1 - 1.2^{11})}{1 - 1.2} = 161 \ \text{(nearest integer)}$$

Example 17

For a geometric series with first term 12 and common ratio $\frac{2}{3}$ find the 5th term, the sum of the first 8 terms and the sum to infinity. Give answers to 3 significant figures where necessary.

$a = 12, \ r = \frac{2}{3}$

5th term is:
$$ar^4 = 12\left(\tfrac{2}{3}\right)^4 = 2.37 \ \text{(3 s.f.)}$$

The sum to 8 terms is:
$$\frac{12[1 - \left(\tfrac{2}{3}\right)^8]}{1 - \tfrac{2}{3}} = 34.6 \ \text{(3 s.f.)}$$

The sum to infinity is:
$$\frac{12}{1 - \tfrac{2}{3}} = 36$$

Example 18

Evaluate $\sum_{n=1}^{7} 16(-\frac{1}{2})^n$ and find the sum to infinity of the series.

The series is $-8 + 4 - 2 + \ldots$ to 7 terms.

This is a geometric series with $a = -8$, $r = -\frac{1}{2}$ and $n = 7$.

The sum to 7 terms is: $\dfrac{(-8)[1 - (-\frac{1}{2})^7]}{1 - (-\frac{1}{2})} = \dfrac{-129}{24} = -5.375$

The sum to infinity is: $\dfrac{a}{1-r} = \dfrac{-8}{1 - (-\frac{1}{2})} = -\dfrac{16}{3} = -5\frac{1}{3}$

Example 19

Find the least number of terms of the geometric series with 1st term 50 and 2nd term 47 for which the sum exceeds 800.

$$a = 50, \ ar = 47 \quad \text{and so} \quad r = \frac{47}{50} = 0.94$$

The first term is 50 and the common ratio is 0.94.

Using $S_n = \dfrac{a(1 - r^n)}{1 - r}$ the sum to n terms is

$$S_n = \frac{50[1 - 0.94^n]}{1 - 0.94}$$

and you require the smallest integer value of n for which $S_n > 800$. That is,

$$1 - 0.94^n > \frac{800 \times 0.06}{50} = 0.96$$

which gives $0.94^n < 0.04$.

Using a method of trial and improvement and a calculator you will find that

$$0.94^{52} = 0.040\,054\,2$$

and

$$0.94^{53} = 0.037\,650\,9$$

This gives the value of n required as 53.

Note: Book P2 shows you another method of solving equations of the type $a^x = b$ using logarithms in place of the trial and improvement approach used here.

Exercise 6E

Where necessary, give answers to 3 s.f.

1 Find the 10th term and the sum of the first 10 terms of the geometric series with first term u_1 and common ratio r when:

   (a) $u_1 = 10$ and $r = 2$

   (b) $u_1 = 10$ and $r = -2$

   (c) $u_1 = 10$ and $r = \frac{1}{2}$

2 Find the sum of the first 12 terms of the geometric series $u_1 + u_2 + u_3 + \ldots + u_{12}$, when:

   (a) $u_1 = 3$, $u_2 = 6$

   (b) $u_1 = 3$, $u_2 = -6$

   (c) $u_1 = 3$, $u_2 = 2$

   (d) $u_1 = 3$, $u_2 = -2$

3 A geometric series has first term a and common ratio r. Find the nth term and the sum to n terms when:

   (a) $a = 4$, $r = \frac{1}{2}$ and $n = 8$

   (b) $a = -4$, $r = 2$ and $n = 10$

   (c) $a = 48$, $r = -\frac{1}{3}$ and $n = 8$

   (d) $a = -100$, $r = 0.9$ and $n = 13$

4 State which series in question 3 has no sum to infinity and explain why.

   Find the sum to infinity of the three convergent series in question 3.

5 Find the first three terms of a geometric series which has a 4th term of -3 and a 7th term of 81. Find also the sum of the first 12 terms of this series.

6 Write down the sum of the first n terms of the following geometric series and find the sum to infinity in those cases where the series is convergent.

   (a) $\frac{3}{10} + \frac{3}{100} + \frac{3}{1000} + \ldots$

   (b) $16 - 8 + 4 - \ldots$

   (c) $8 - 12 + 18 - \ldots$

7 The sum to infinity of a geometric series is 5 and the first term is 7. Find the common ratio and the sum of the first 15 terms.

8   The sum to infinity of a geometric series is 9. The common ratio is positive and the sum of the first two terms is 5. Find the first term, the common ratio and the sum of the first 12 terms.

9   A geometric series has the following properties. The 1st and 2nd terms have a sum of -4 and the 4th and 5th terms have a sum of 108. Find the 1st term and the common ratio of the series. Explain why the series has no sum to infinity.

10 A state's population at the end of each year is 2% greater than at the beginning of the year. Find the number of years required for the population to double.

11 The first term of a geometric series is 500 and the common ratio is 0.93. The nth term of the series is u_n. Find the least value of n for which (a) $u_n < 40$ (b) $u_n < 30$.

12 The 1st and the 4th terms of a geometric series are 27 and 8 respectively. Find the 7th term and the common ratio. Find also the sum to infinity.

13 By the end of a year, the value of a shop has increased by 3% of its value at the start of the year. At the start of 1994 a shop was valued at £45 000. Estimate, to the nearest £100, the value of the shop at the end of the year (a) 2000 (b) 2200.

14 Find to 3 significiant figures the values of:

(a) $\displaystyle\sum_{n=4}^{9}(1.5)^n$ (b) $\displaystyle\sum_{n=2}^{10}4(\tfrac{3}{4})^{n-1}$

15 Find the value of:

$$\sum_{n=1}^{10}[3(\tfrac{2}{3})^n - 2]$$

giving your answer to 3 significant figures.

6.7 Further examples on arithmetic and geometric series

Example 20

Evaluate $\displaystyle\sum_{n=1}^{11}(1.2^n + 1.2n)$.

You can make this easier by separating it into two summations:

$$\sum_{n=1}^{11} 1.2^n + \sum_{n=1}^{11} 1.2n$$

The first is a geometric series with first term 1.2 and common ratio 1.2 which has a sum of

$$\frac{1.2(1 - 1.2^{11})}{1 - 1.2} = 38.58$$

The second is an arithmetic series with first term 1.2 and common difference 1.2 which has a sum of

$$\frac{11}{2}[2.4 + 10 \times 1.2] = 79.2$$

Hence the sum of the series is $79.2 + 38.58 = 117.78$.

Example 21

The 1st, 2nd and 3rd terms of an arithmetic series are a, b and a^2, where a is negative. The 1st, 2nd and 3rd terms of a geometric series are a, a^2 and b. Find: (a) the values of a and b (b) the sum to infinity of the geometric series (c) the sum of the first 40 terms of the arithmetic series.

(a)   As a, b and a^2 are consecutive terms in an arithmetic series,

$$b - a = a^2 - b$$

which gives:

$$2b = a + a^2 \qquad \text{(i)}$$

As a, a^2 and b are consecutive terms in a geometric series,

$$\frac{a^2}{a} = \frac{b}{a^2}$$

which gives:

$$a^3 = b \qquad \text{(ii)}$$

Eliminating b from equations (i) and (ii) gives:

$$2a^3 = a^2 + a$$

Since a is not zero,

$$2a^2 - a - 1 = 0$$

Factorising: $(2a + 1)(a - 1) = 0$

It follows, since a is negative, that $a = -\frac{1}{2}$ and $b = -\frac{1}{8}$.

(b)   The first term of the geometric series is $-\frac{1}{2}$ and the common ratio is $-\frac{1}{2}$.

The sum to infinity $= \dfrac{-\frac{1}{2}}{1-(-\frac{1}{2})} = -\frac{1}{3}$

(c) The arithmetic series has first term $-\frac{1}{2}$ and common difference $(-\frac{1}{8} + \frac{1}{2}) = \frac{3}{8}$.

The sum of the first 40 terms $= 20[-1 + 39(\frac{3}{8})]$

$$= 272.5$$

Exercise 6F

1 The numbers p, 10 and q are three consecutive terms of an arithmetic series. The numbers p, 6 and q are three consecutive terms of a geometric series. Show that $p^2 - 20p + 36 = 0$ and hence find the values of p and of q for which the geometric series converges.

Find the sum to infinity of the geometric series in this case.

2 A geometric series has all terms positive. The sum of the first four terms is 15 and the sum to infinity of the series is 16. Find the sum of the first eight terms.

3 A ball is dropped from a height of 10 m and bounces on a horizontal floor to a height of 8 m. On each successive bounce the height reached is 0.8 times the height reached in the previous bounce. Find the total distance travelled by the ball before it comes to rest.

4 The first term of an arithmetic series is -10.5 and the common difference is 1.5. The sum of n terms is -42. Find the two values of n which satisfy these conditions.

5 A geometric series has 1st term $\sqrt{2}$ and 2nd term $\sqrt{6}$. Find the 12th term and the sum of the first 12 terms.

6 The sum of n terms of an arithmetic series is 36. The first term is 1 and the nth term is 11. Find the value of n and the common difference of the series.

7 The first three terms of a geometric series are $4x$, $x + 1$ and x. Given that x is negative, find the sum to infinity of the series.

8 For the arithmetic series $9.5 + 9.1 + \ldots$, find:
   (a) the 50th term
   (b) the least value of n for which the sum of the first n terms is negative.

9 A geometric series has 4th term 10 000 and 8th term 1. Find two series which satisfy these data and find the sum to infinity of each of these series.

10 Evaluate:

$$\sum_{n=2}^{9} [(0.9)^n + (0.9n + 4)]$$

giving your answer to 2 decimal places.

11 The sum of the first n terms of a series is $(4n + 5)^2$. Find the nth term of the series.

SUMMARY OF KEY POINTS

1 A sequence is a succession of terms

$$u_1, u_2, u_3, \ldots$$

that are formed by a rule or a recurrence relation.

2 A sequence may approach a limiting value, it may oscillate, it may be periodic or it may diverge.

3 A sequence that ends after a finite number of terms is called a finite sequence. A sequence that continues infinitely is called an infinite sequence.

4 A series is the sum of the terms in a sequence:

$$S_n = u_1 + u_2 + u_3 + \ldots + u_n = \sum_{r=1}^{n} u_r$$

This series has a finite number of terms and is therefore called a finite series. The Greek letter Σ is known as the summation sign. Its lower and upper limits show the values of the variable r over which the summation is being made.

5 An infinite series occurs when the variable n has no upper limit; the series then has an unlimited number of terms.

6 The arithmetic series:

$$a + (a + d) + (a + 2d) + \ldots + [a + (n - 1)d]$$

has n terms. Its sum is:

$$\frac{n}{2}[2a + (n - 1)d]$$

For an arithmetic series with n terms, whose first term is a and last term L, the sum is:

$$\frac{n}{2}\left[a + L\right]$$

7 The geometric series:

$$a + ar + ar^2 + \ldots + ar^{n-1}$$

has n terms. The sum is:

$$\frac{a(1 - r^n)}{1 - r} \quad \text{or} \quad \frac{a(r^n - 1)}{r - 1}$$

8 The infinite geometric series:

$$a + ar + ar^2 + \ldots$$

has a sum to infinity of $\dfrac{a}{1 - r}$, provided that $|r| < 1$. In this case, the series is said to converge.

Review exercise

1 Find the equation of the line joining (2,7) to (5,3).

2 Find, without using a calculator the numerical values of
   (a) $25^{\frac{1}{2}}$, (b) 3^{-2}, (c) $27^{-\frac{1}{3}}$, (d) 6^0. [L]

3 Find the positive constants a and b such that 0.25, a, 9 are in geometric progression and 0.25, a, $9 - b$ are in arithmetic progression. [L]

4 Show that the triangle with vertices at $(-3, -5)$, $(2,5)$ and $(8,2)$ is right-angled.

5 Find, without using a calculator, the numerical values of
   (a) $\left(\dfrac{8}{27}\right)^{\frac{1}{3}}$, (b) 4^{-2}, (c) $16^{\frac{3}{2}}$, (d) 7^0. [L]

6 Given that

$$S_n = \sum_{r=1}^{n} 5^r,$$

   where n is a positive integer, show that

$$S_n = \tfrac{5}{4}(5^n - 1).$$

   Find the least value of n for which

$$S_n > 2^{50}$$ [L]

7 The points $P(2,p)$ and $Q(q+1, 3q-2)$ both lie on the line whose equation is $y = 5x + 1$. Find
   (a) the values of p and q,
   (b) the distance, to 3 significant figures, between the points P and Q. [L]

8 Write down the next two terms in the sequence:
   (a) $\frac{1}{3}$, $\frac{1}{6}$, $\frac{1}{12}$, $\frac{1}{24}$, $\cdots$
   (b) 1, $\frac{2}{3}$, $\frac{3}{9}$, $\frac{4}{27}$, $\cdots$
   (c) $1, 2, 6, 24, \ \cdots$

9 Given that $x = 16$, find, without using a calculator, the value of

 (a) $x^{\frac{1}{2}}$, (b) $x^{\frac{3}{2}}$, (c) $x^{-\frac{1}{4}}$. [L]

10 The first three terms of a geometric series are $3(q + 5)$, $3(q + 3)$ and $(q + 7)$ respectively.

 (a) Calculate the value of q.

 (b) Find the common ratio, r, of the geometric series.

 (c) Find the sum to infinity of the geometric series. [L]

11 The points A, B and C have coordinates $(6, -9)$, $(-1, 15)$ and $(-10, 3)$ respectively. Prove that $\angle BCA = 90°$. Hence calculate the cosine of $\angle CBA$. [L]

12 Without using a calculator, write down the value of

 (a) $(37)^0$, (b) $8^{\frac{1}{3}}$, (c) $3^2 \times (27)^{-\frac{2}{3}}$, (d) $(-\frac{1}{2})^{-2}$. [L]

13 Without using a calculator, solve the equation

$$3^{(2x-1)} = 9^{-x}.$$ [L]

14 (a) A geometric series has first term 3 and common ratio 0.8. Find the sum of the first 24 terms, giving your answer to 3 significant figures.

 (b) Find $\displaystyle\sum_{r=1}^{52} \frac{r}{2}$. [L]

15 Show that the triangle with vertices at $(5, 9)$, $(2, 3)$ and $(8, 6)$ is isosceles.

16 Solve the equation

$$4(2^{x^2}) = 16^x.$$ [L]

17 John is given an interest-free loan to buy a second-hand car. He repays the loan in monthly instalments. He repays £20 the first month, £22 the second month and the repayments continue to rise by £2 per month until the loan is repaid. Given that the final monthly repayment is £114,

 (a) show that the number of months it will take John to repay the loan is 48,

 (b) find the amount, in pounds, of the loan. [L]

18 Write down the first six terms of the sequence defined by

$$u_n - u_{n-1} - 2u_{n-2} = 0, \; u_0 = 4, \; u_1 = -1$$

19 With respect to the origin O, the vertices of $\triangle ABC$ are $A(2, 5)$, $B(2, -1)$ and $C(-2, 3)$.

(a) Prove that for all values of t, the points with coordinates $(t - 1, t)$ are equidistant from B and C.

(b) Given that the point D is equidistant from A, B and C, calculate the coordinates of D. [L]

20 Find, without using a calculator, the value of

(a) 9^{-2}, (b) $27^{\frac{1}{3}}$, (c) $4^{2.5}$, (d) x for which $7^x = 1$. [L]

21 (i) Find $\displaystyle\sum_{r=1}^{100} \frac{2r}{3}$.

(ii) Find the sum to infinity of

$$\frac{7}{10} + \frac{7}{100} + \frac{7}{1000} + \frac{7}{10\,000} + \ldots,$$

giving your answer in the form $\dfrac{p}{q}$, where p and q are positive integers.

Hence, or otherwise, find the value of

$\displaystyle\sum_{r=1}^{\infty} \frac{k}{10^r}$ in terms of k. [L]

22 The line through $(2, 5)$ with gradient 3 cuts the x-axis at A and the y-axis at B. Calculate the area of $\triangle AOB$.

23 Solve the equation $81^x + 81 = 30(9^x)$. [L]

24 An investment of £2000 is made at the start of a year with a Finance Company. At the end of this year and at the end of each subsequent year the value of the investment is 11% greater than its value at the start of that year.

(a) Find, to the nearest £, the value of the investment at the end of

(i) the 5th year,

(ii) the 10th year.

A client decides to invest £2000 at the start of each year. Write down a series whose sum is the total value of this annual investment at the end of 12 years.

(b) By finding the sum of your series, determine, to the nearest £, the value of the investment at the end of 12 years. [L]

25 Solve the equation $5(2^x) - 4^x = 4$. [L]

26 A ball is dropped onto a horizontal plane and rebounds successively. The height above the plane reached by the ball after the first impact with the plane is $4\,\text{m}$. After each impact the ball rises to a height which is $\frac{3}{4}$ of the height reached after the previous impact. Calculate the total vertical distance travelled by the ball from the first impact until

(a) the fourth impact,

(b) the eighth impact,

(c) it comes to rest. [L]

27 Find, without using a calculator, the numerical value of

(a) $8^{\frac{1}{3}}$, (b) 10^{-4}, (c) 49^0, (d) $\dfrac{8^{\frac{1}{2}} \times 2^{\frac{1}{2}}}{16^{\frac{1}{4}}}$ [L]

28 The line with equation $y = x + 2$ intersects the curve with equation $x^2 + y^2 = 12 - 2x$ at the points A and B. The coordinates of A are both positive. Calculate the coordinates of (a) the point A, (b) the point B. Hence calculate the length AB. [L]

29 Without using a calculator, solve the equation

$$4^{2x-1} = 16^{-\frac{1}{2}x}$$ [L]

30 (i) The 3rd term of an arithmetic series is -20 and the 11th term is 20. Find the common difference and the 1st term of this series.

When the first k terms of this series are added together, their sum is zero. Find the non-zero value of k.

(ii) The 1st and 7th terms of a geometric series of positive terms are 2 and 16 respectively.

Find, to 3 significant figures, the sum of the first 7 terms of this series. [L]

31 Given that

$$u_{n+2} = \tfrac{1}{2}(u_{n+1} + u_n), \quad u_1 = 2, \; u_2 = 5$$

write down the first seven terms in the sequence defined by the recurrence relation.

32 A circle has its centre at $(3, -4)$ and has radius 13. Show that $(-2, 8)$, lies on the circumference of the circle.

33 Without using a calculator, find the value of

$$4^{\frac{1}{4}} \times 2^3 \times 4^{\frac{3}{4}}.$$ [L]

34 (i) $$S_n - \sum_{r=1}^{n} (2r - 3).$$

(a) Write down the first three terms of this series.

(b) Find S_{50}.

(c) Find n such that $S_n = 575$.

(ii) The first three terms of a geometric series are $(x - 8)$, x and $(2x + 12)$ respectively.

(a) Find the two possible values of x and, hence, the two possible values of the common ratio of the series.

Given also that the common ratio is less than one,

(b) find the sum to infinity of the series.   [L]

35 The line with equation $y = 3x + 1$ cuts the curve with equation $y = x^2 + 4x - 5$ at the points A and B. Calculate the length AB.

36 Given that $27^x = 9^{x-1}$, find the value of x.   [L]

37 The first three terms of a series, S, are $(m - 4)$, $(m + 2)$ and $(3m + 1)$.

(i) Given, also, that S is an arithmetic series,

(a) find m.

Using your value of m,

(b) write down the first four terms of the arithmetic series.

(ii) Given, instead, that S is a geometric series,

(a) find the two possible values of m,

(b) write down the first four terms of each of the two geometric series obtained with your values of m,

(c) state the value of the common ratio of each of the series.

One of these two geometric series has a sum to infinity.

(d) Find the sum to infinity of that series.   [L]

38 The straight line through the points $(1,4)$ and $(-3, -4)$ meets the coordinate axes in the points A and B. Find the area of a square having AB as one of its sides.   [L]

39 (i) Given that $x = 27$, find, without using a calculator, the values of

(a) $x^{\frac{1}{3}}$, (b) $x^{-\frac{2}{3}}$, (c) x^0.

(ii) Given that $2^{2x-1} = \frac{1}{8}$, find the value of x. [L]

40 (i) In a geometric series, $(x - 1)$, $(x + 1)$ and $(x + 9)$ are consecutive terms. Find the value of the common ratio of the series.

(ii) The sum of the first n terms of an arithmetic series is 6014. Given that the sum of the first term and the nth term is 124,

(a) calculate the value of n,

(b) show that the value of the 49th term is 62.

Given also that the value of the 61st term is 77, find

(c) the value of the first term of the series,

(d) the common difference,

(e) the sum of the first 80 terms. [L]

41 The centre of a circle, radius 2, is at $(1,2)$ and a diameter has equation $y = 2x$.

Calculate, to 2 significant figures the x-coordinates of the ends of this diameter.

42 Simplify as far as possible

$$\frac{9a^3b^2}{2c^4} \div \frac{3a^2b^4}{4ac^2}$$ [L]

43 The third term of a geometric series is 15 and the common ratio of the series is 2.

(a) Calculate the value of the sixth term of the series.

(b) Calculate the sum of the first ten terms of the series.

The second and fifth terms of the series form the first two terms of an arithmetic series.

(c) Find the value of the ninth term of the arithmetic series.

(d) Find the sum of the first thirteen terms of the arithmetic series. [L]

44 Find the area of the triangle with vertices at $(2,1)$, $(5,3)$ and $(4,6)$.

45 Find, without using a calculator, the numerical values of

(a) $2^5 - 2^3$, (b) $32^{\frac{3}{5}}$, (c) $8^{-\frac{1}{3}}$. [L]

46 Of the three series

(i) $1 - \frac{1}{3} + \frac{1}{9} - \frac{1}{27} + \dots$,

(ii) $1 - \frac{1}{3} - 1 - \frac{4}{3} - \dots$,

(iii) $1 - \frac{1}{3} - 1\frac{2}{3} - 3 - \dots$,

one is an arithmetic series and one is a geometric series.

(a) Find the tenth term of the arithmetic series.

(b) Find the sixth term of the geometric series.   [L]

47 Calculate the coordinates of the point of intersection of the line joining $(5, -4)$ to $(-3, 4)$ with the line joining $(-13, 4)$ to $(-5, 1)$.

48 (a) Write each of the following in the form x^n, where n is an integer.

(i) $\sqrt{x^2}$,   (ii) $(x^2)^4$,   (iii) $\dfrac{x^2}{x^8}$.

(b) Find the value of $64^{\frac{2}{3}}$.   [L]

49 (i) The first and third terms of an arithmetic series are $\frac{2}{3}$ and $\frac{3}{2}$ respectively. Find

(a) the common difference of this series,

(b) the eighth term of this series.

(ii) Two distinct geometric series each have first term $\frac{2}{3}$ and third term $\frac{3}{2}$.

(a) Find the common ratio of each series.

(b) Use the formula

$$S_n = \frac{a(r^n - 1)}{r - 1} ,$$

where S_n is the sum of the first n terms of a geometric series with first term a and common ratio r, to find the sum, to 2 significant figures, of the first 8 terms of each series.   [L]

50 Write out the following series and evaluate them:

(a) $\displaystyle\sum_{r=1}^{5} r(r + 1)$ (b) $\displaystyle\sum_{r=2}^{5} \frac{1}{r(r - 1)}$

51 The vertices of $\triangle ABC$ are the points $A(-1, 5)$, $B(-5, 2)$ and $C(8, -7)$.

(a) Find in the form $px + qy + r = 0$, where p, q and r are integers, an equation of the line passing through B and C.

(b) Show that AB and AC are perpendicular.   [L]

52 Given that $2^{(2x-1)} = \frac{1}{8}$, find the value of x.   [L]

53 (i) The eleventh term of an arithmetic series is 22 and the sum of the first 4 terms is 37.

(a) Find the first term and the common difference of this series.

(b) Find the sum of the first 21 terms of this series.

(ii) The first, second and third terms of a geometric series are x, $(x + 1)$ and $(x + 4)$, respectively.

(a) Find the value of x.

(b) Find the common ratio of the series.

(c) Find the seventh term of the series. [L]

54 Find the coordinates of the point where the line through $(-3, 13)$ and $(6,10)$ cuts the line through $(1,5)$ with gradient 3.

55 (i) The rth term of an arithmetic series is given by

$$2r - 5, \qquad r = 1, 2, 3, \ldots.$$

Write down the first four terms of this series and find

$$S_n = \sum_{r=1}^{n} (2r - 5)$$

in terms of n.

Given that $S_n = 165$ find the value of n.

(ii) In a geometric series, $(x + 1)$, $(x + 3)$ and $(x + 4)$ are the first, second and third terms respectively. Calculate the value of x and hence write down the numerical values of the common ratio and the first term of the series.

Calculate the numerical value of the sum to infinity of the series. [L]

56 Find, without using a calculator, the numerical value of

(a) $144^{\frac{1}{2}}$, (b) 3^{-3}, (c) $3^{\frac{3}{4}} . 3^{-\frac{1}{2}} . 3^{-\frac{1}{4}}$. [L]

57 (i) The seventh term of an arithmetic series is 6 and the eighteenth term is $22\frac{1}{2}$. Calculate

(a) the common difference of the series,

(b) the first term of the series.

Given also that the sum of the first n terms of the series is 252,

(c) find the value of n.

(ii) The second term of a convergent geometric series is 24 and the sum of the first three terms is 126. Find the common ratio of the series. [L]

58 Write down the first seven terms in the sequence defined by

$$u_{r+2} = u_{r+1} + 6u_r, \; u_1 = -1, \; u_2 = 1$$

Trigonometry

7

7.1 Measuring angles in radians

In GCSE mathematics courses angles are usually measured in units called degrees, where $360°$ is a complete circle. There is no real need for a complete revolution to be divided into 360 equal parts – this is just a convention. A revolution could just as well be divided into 100 parts, 1000 parts or any other number of equal parts and a suitable name given to the new unit.

In the study of trigonometry and in mathematics generally you will often find it useful to measure the size of an angle in units called **radians**.

- **One radian is the angle subtended at the centre by the arc of a circle whose length is equal to the radius of the circle.**

'The angle subtended by the arc of a circle' just means the angle between the radii that join each end of the arc to the centre of the circle.

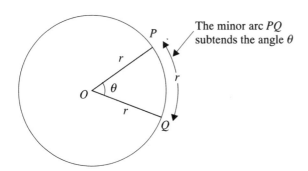

In the diagram the radius of the circle is r and the length of the arc PQ is also r. As the arc is the same length as the radius of the circle, the angle θ is 1 radian. This is often abbreviated to 1^c or $1\,\text{rad}$.

The angle made at the centre of the circle by an arc of length r is 1^c. The circumference of the circle is $2\pi r$ which is 2π times the length of this arc. So the angle of one revolution is $2\pi \times 1^c = 2\pi^c$ (two pi radians).

One full turn is 360 degrees. One full turn in radians is 2π radians. So:

$$360° = 2\pi^c$$

Halving gives:

$$180° = \pi^c$$

and halving again:

$$90° = \frac{\pi^c}{2}$$

Example 1
Convert 60° to radians.

$$360° = 2\pi^c$$

$$1° - \frac{2\pi^c}{360}$$

$$60° = \frac{2\pi^c}{360} \times 60$$

$$= \frac{\pi^c}{3}$$

Example 2
Convert $\dfrac{5\pi^c}{6}$ to degrees.

$$2\pi^c = 360°$$

$$1^c = \frac{360°}{2\pi}$$

$$\frac{5\pi^c}{6} = \frac{360}{2\pi} \times \frac{5\pi}{6} \text{ degrees}$$

$$= 150°$$

7.2 Finding the length of an arc of a circle

Suppose that the arc PQ of a circle subtends an angle θ radians at the centre of the circle. As the complete angle at the centre of the circle is of size 2π radians:

$$\frac{\text{length of arc } PQ}{\text{circumference of circle}} = \frac{\theta}{2\pi}$$

The circumference of the circle is $2\pi r$. If we say the length of the arc PQ is s then:

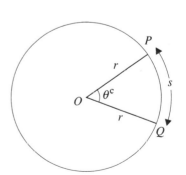

$$\frac{s}{2\pi r} = \frac{\theta}{2\pi}$$

So: $$s = \frac{\theta}{2\pi} \times 2\pi r$$

and: $$s = r\theta$$

- **In general the length s of an arc of a circle is: $s = r\theta$, where r is the radius and θ is the angle in radians subtended by the arc at the centre of the circle.**

Example 3

Find the length of the minor arc of a circle of radius 8 cm, given that the arc subtends an angle of 2 radians at the centre of the circle.

$s = r\theta = 8 \times 2\,\text{cm} = 16\,\text{cm}$

Example 4

An arc of a circle of length 5 cm subtends an angle of $1\frac{1}{2}$ radians at the centre of the circle. Calculate the radius of the circle.

$$s = r\theta$$
$$5 = r \times 1\frac{1}{2}$$
$$r = \frac{5}{1\frac{1}{2}} = 3\frac{1}{3}$$

The radius is $3\frac{1}{3}$ cm.

7.3 Finding the area of a sector of a circle

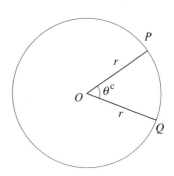

You can find the area of the sector OPQ in a similar way to that in which you found the length of the arc PQ.

$$\frac{\text{area of sector } OPQ}{\text{area of circle}} = \frac{\theta}{2\pi}$$

The area of the circle is πr^2 so if we call the area of the sector A, then:

$$\frac{A}{\pi r^2} = \frac{\theta}{2\pi}$$

$$A = \frac{\theta}{2\pi} \times \pi r^2$$

$$A = \tfrac{1}{2}r^2\theta$$

■ **In general, the area A of a sector of a circle is:**

$$A = \tfrac{1}{2}r^2\theta$$

where r is the radius and θ is the angle in radians subtended by the arc at the centre of the circle.

Example 5

Find the area of the minor sector OPQ of a circle, given that the arc PQ subtends an angle 1.2 radians at the centre of the circle and that the radius of the circle is 5 cm.

$$A = \tfrac{1}{2}r^2\theta$$

$$= \tfrac{1}{2} \times 5^2 \times 1.2 \, \text{cm}^2$$

$$= 15 \, \text{cm}^2$$

Example 6

In the diagram the area of the minor sector OPQ is $18 \, \text{cm}^2$. The angle POQ is of size 0.64 radians. Calculate the radius of the circle.

$$A = \tfrac{1}{2}r^2\theta$$

$$18 = \tfrac{1}{2} \times r^2 \times 0.64$$

$$r^2 = \frac{18}{0.32} = 56.25$$

$$r = 7.5$$

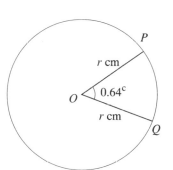

The radius of the circle is 7.5 cm.

Exercise 7A

1 Convert the following to degrees:

(a) π^c (b) $\dfrac{\pi^c}{2}$ (c) $\dfrac{\pi^c}{6}$ (d) $\dfrac{\pi^c}{4}$ (e) $\dfrac{3\pi^c}{2}$

(f) $\dfrac{3\pi^c}{4}$ (g) $4\pi^c$ (h) $3\pi^c$ (i) $\dfrac{5\pi^c}{4}$ (j) $\dfrac{5\pi^c}{6}$

2 Convert the following to radians, leaving your answer in terms of π:

(a) $22\frac{1}{2}°$ (b) $15°$ (c) $180°$ (d) $210°$ (e) $120°$

(f) $135°$ (g) $225°$ (h) $330°$ (i) $300°$ (j) $315°$

3 Convert the following to degrees, giving your answer to the nearest degree:

(a) 1^c (b) 2.4^c (c) 5^c (d) 3.5^c (e) 0.4^c

(f) 1.7^c (g) 4.3^c (h) 5.5^c (i) 4^c (j) 6^c

4 Convert the following to radians, giving your answer to two decimal places:

(a) $20°$ (b) $50°$ (c) $70°$ (d) $130°$ (e) $170°$

(f) $230°$ (g) $250°$ (h) $85°$ (i) $38°$ (j) $152°$

5 Find the length of the minor arc PQ and the area of the corresponding sector in each case:

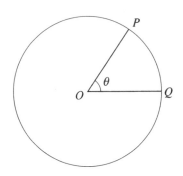

(a) $OP = 3\,\text{cm}$, $\theta = 1^c$ (b) $OP = 7\,\text{cm}$, $\theta = 2^c$

(c) $OP = 8\,\text{cm}$, $\theta = 1.5^c$ (d) $OP = 9.5\,\text{cm}$, $\theta = 3^c$

(e) $OP = 5.5\,\text{cm}$, $\theta = 0.5^c$ (f) $OP = 9\,\text{cm}$, $\theta = 2.8^c$

(g) $OP = 11\,\text{cm}$, $\theta = 4.5^c$ (h) $OP = 9.5\,\text{cm}$, $\theta = 5^c$

(i) $OP = 11.2\,\text{cm}$, $\theta = 4.7^c$ (j) $OP = 8.3\,\text{cm}$, $\theta = 6.1^c$

6 The arc PQ of a circle has length $15\,\text{cm}$ and the radius of the circle is $6\,\text{cm}$. Find the angle, in radians, subtended by the arc at the centre of the circle.

7 An arc is of length $9\,\text{cm}$ and it subtends an angle of 2.5 radians at the centre of a circle. Calculate the radius of the circle.

8 Given that the area of the minor sector OPQ in the diagram for question 5 is $10\,\text{cm}^2$ and that $OP = 4\,\text{cm}$, calculate θ in radians.

9 Given that the area of the minor sector OPQ is $50\,\text{cm}^2$ and that $\theta = 1.4$ radians, calculate the radius of the circle.

10 Calculate the area of the shaded region in the diagram, given that:

(a) $OC = 1\,\text{cm}$, $CA = 1\,\text{cm}$, $\theta = 2^c$

(b) $OC = 1\,\text{cm}$, $CA = 0.5\,\text{cm}$, $\theta = 1.5^c$

(c) $OC = 1.5\,\text{cm}$, $CA = 0.5\,\text{cm}$, $\theta = 2.5^c$

(d) $OC = 2\,\text{cm}$, $CA = 2.5\,\text{cm}$, $\theta = 0.7^c$

(e) $OC = 2.7\,\text{cm}$, $CA = 1.2\,\text{cm}$, $\theta = 1.6^c$

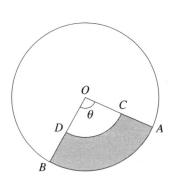

7.4 The three basic trigonometric functions for acute angles

For any right-angled triangle, the three basic trigonometric functions are defined as:

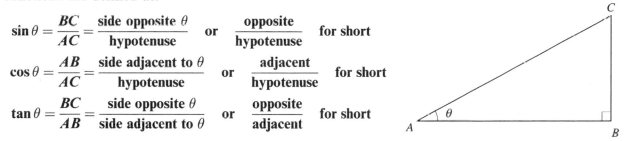

$$\sin\theta = \frac{BC}{AC} = \frac{\text{side opposite } \theta}{\text{hypotenuse}} \quad \text{or} \quad \frac{\text{opposite}}{\text{hypotenuse}} \quad \text{for short}$$

$$\cos\theta = \frac{AB}{AC} = \frac{\text{side adjacent to } \theta}{\text{hypotenuse}} \quad \text{or} \quad \frac{\text{adjacent}}{\text{hypotenuse}} \quad \text{for short}$$

$$\tan\theta = \frac{BC}{AB} = \frac{\text{side opposite } \theta}{\text{side adjacent to } \theta} \quad \text{or} \quad \frac{\text{opposite}}{\text{adjacent}} \quad \text{for short}$$

These definitions must be remembered.

Example 7

From your calculator write down to 3 significant figures the value of:
(a) $\sin 70.1°$ (b) $\cos 47.3°$ (c) $\tan 37.8°$ (d) $\cos 81.7°$ (c) $\tan 52.9°$

(a) $\sin 70.1° = 0.940\,288\,1 = 0.940$ (3 s.f.)

(b) $\cos 47.3° = 0.678\,159\,6 = 0.678$ (3 s.f.)

(c) $\tan 37.8° = 0.775\,679\,5 = 0.776$ (3 s.f.)

(d) $\cos 81.7° = 0.144\,356\,2 = 0.144$ (3 s.f.)

(e) $\tan 52.9° = 1.322\,237 \ = 1.32$ (3 s.f.)

Example 8

From your calculator write down to 3 significant figures the value of:
(a) $\sin 0.13^c$ (b) $\cos 0.62^c$ (c) $\tan 1.13^c$ (d) $\sin 0.891^c$ (e) $\cos 0.473^c$

(a) $\sin 0.13^c = 0.129\,634\,1 = 0.130$ (3 s.f.)

(b) $\cos 0.62^c = 0.813\,878\,4 = 0.814$ (3 s.f.)

(c) $\tan 1.13^c = 2.119\,750\,1 = 2.12$ (3 s.f.)

(d) $\sin 0.891^c = 0.777\,700\,7 = 0.778$ (3 s.f.)

(e) $\cos 0.473^c = 0.890\,205\,6 = 0.890$ (3 s.f.)

If you can use a calculator to find the value of $\sin\theta$, $\cos\theta$ or $\tan\theta$, where θ is acute, then you can also reverse the process and find θ, if it is acute, from the value of $\sin\theta$, $\cos\theta$ or $\tan\theta$. The inverse functions to sin, cos and tan are called **arcsin**, **arccos** and **arctan**. There should be inverse buttons that allow you to use these functions on your calculator. Notice that on most calculators arcsin θ appears as $\sin^{-1}\theta$, arccos θ as $\cos^{-1}\theta$ and arctan θ as $\tan^{-1}\theta$. However this book uses the notation arcsin θ and so on.

For example, if $\sin\theta = 0.7$, write $\theta = \arcsin 0.7$ and find this using your calculator. Similarly, if $\cos\theta = 0.7$ write $\theta = \arccos 0.7$ and find this using your calculator. The same applies to $\tan\theta = 0.7$ which we can write as $\theta = \arctan 0.7$.

Example 9

Find, in degrees to one decimal place, the acute angle such that:
(a) $\sin\theta = 0.7976$ (b) $\cos\theta = 0.3391$ (c) $\tan\theta = 1.5923$
(d) $\sin\theta = 0.8824$ (e) $\cos\theta = 0.1125$.

(a) $\sin\theta = 0.7976$

So:
$$\theta = \arcsin 0.7976$$
$$= 52.901\,52..^{\circ}$$
$$= 52.9^{\circ} \ (1\,\text{d.p.})$$

(b) $\cos\theta = 0.3391$

So:
$$\theta = \arccos 0.3391$$
$$= 70.177\,94..^{\circ}$$
$$= 70.2^{\circ} \ (1\,\text{d.p.})$$

(c) $\tan\theta = 1.5923$

So:
$$\theta = \arctan 1.5923$$
$$= 57.870\,260..^{\circ}$$
$$= 57.9^{\circ} \ (1\,\text{d.p.})$$

(d) $\sin\theta = 0.8824$

So:
$$\theta = \arcsin 0.8824$$
$$= 61.933\,243..^{\circ}$$
$$= 61.9^{\circ} \ (1\,\text{d.p.})$$

(e) $\cos\theta = 0.1125$

So:
$$\theta = \arccos 0.1125$$
$$= 83.540\,550..^{\circ}$$
$$= 83.5^{\circ} \ (1\,\text{d.p.})$$

Example 10
Find in radians to 3 significant figures the acute angle such that:
(a) $\tan \theta = 2.0156$ (b) $\cos \theta = 0.1738$ (c) $\sin \theta = 0.7014$
(d) $\tan \theta = 1.5721$ (e) $\sin \theta = 0.3333$

(a) $\tan \theta = 2.0156$

So.
$$\theta = \arctan 2.0156$$
$$= 1.110\,249..^{c}$$
$$= 1.11^{c} \ (3\,\text{s.f.})$$

(b) $\cos \theta = 0.1738$

So:
$$\theta = \arccos 0.1738$$
$$= 1.396\,109\,2..^{c}$$
$$= 1.40^{c} \ (3\,\text{s.f.})$$

(c) $\sin \theta = 0.7014$

So:
$$\theta = \arcsin 0.7014$$
$$= 0.777\,359..^{c}$$
$$= 0.777^{c} \ (3\,\text{s.f.})$$

(d) $\tan \theta = 1.5721$

So:
$$\theta = \arctan 1.5721$$
$$= 1.004\,260..^{c}$$
$$= 1.00^{c} \ (3\,\text{s.f.})$$

(e) $\sin \theta = 0.3333$

So:
$$\theta = \arcsin 0.3333$$
$$= 0.339\,801\,5..^{c}$$
$$= 0.340^{c} \ (3\,\text{s.f.})$$

Example 11
Given that θ is acute and $\tan \theta = 1.1714$, work in degrees to find, to
3 significant figures, the value of $\sin \theta$.

$\tan \theta = 1.1714$

So:
$$\theta = \arctan 1.1714$$
$$= 49.513\,298..^{\circ}$$

and:
$$\sin \theta = \sin 49.513\,298..^{\circ}$$
$$= 0.760\,556\,68..$$
$$= 0.761 \ (3\,\text{s.f.})$$

Notice that in questions such as this you should leave the full display, $49.513\,298\ldots$, in the calculator before pressing the sin button. Do not approximate the angle to 3 s.f. first, or you will introduce approximation errors.

Example 12

Working in radians find to 3 significant figures the value of $\cos\theta$, given that $\sin\theta = 0.321\,46$.

$$\sin\theta = 0.321\,46$$
$$\theta = \arcsin 0.321\,46$$
$$= 0.327\,270\ldots\text{ (radians)}$$
$$\cos\theta = 0.946\,923\,1\ldots$$
$$= 0.947\ (3\text{ s.f.})$$

Exercise 7B

1 Find to 3 significant figures the value of:
   (a) $\sin 28.1°$ (b) $\cos 63.2°$
   (c) $\tan 57.3°$ (d) $\cos 31.4°$
   (e) $\tan 48.4°$ (f) $\sin 15.7°$
   (g) $\tan 23.8°$ (h) $\cos 44.4°$
   (i) $\cos 77.3°$ (j) $\tan\ 43.4°$

2 Find to 3 significant figures the value of:
   (a) $\sin 0.2317^c$ (b) $\cos 0.341\,25^c$
   (c) $\tan 0.991\,956^c$ (d) $\sin 1.001\,23^c$
   (e) $\cos 0.124\,678^c$ (f) $\tan 0.314\,26^c$
   (g) $\tan 1.123\,214^c$ (h) $\sin 1.001\,314^c$
   (i) $\cos 1.101\,46^c$ (j) $\sin 0.321\,54^c$

3 Find in degrees to 1 decimal place the acute angle such that:
   (a) $\sin\theta = 0.341\,72$ (b) $\cos\theta = 0.112\,16$
   (c) $\tan\theta = 1.521\,34$ (d) $\cos\theta = 0.331\,32$
   (e) $\cos\theta = 0.471\,26$ (f) $\sin\theta = 0.131\,47$
   (g) $\tan\theta = 0.661\,67$ (h) $\sin\theta = 0.781\,23$
   (i) $\tan\theta = 0.531\,32$ (j) $\cos\theta = 0.616\,14$

4 Find in radians to 3 significant figures the acute angle θ such that:
   (a) $\sin\theta = 0.241\,37$ (b) $\cos\theta = 0.441\,22$
   (c) $\tan\theta = 0.891\,92$ (d) $\tan\theta = 1.824\,87$

(e) $\sin\theta = 0.771\,75$ (f) $\cos\theta = 0.662\,26$

(g) $\sin\theta = 0.792\,93$ (h) $\cos\theta = 0.542\,41$

(i) $\tan\theta = 2.1012$ (j) $\cos\theta = 0.241\,47$

5   Given that θ is acute, find to 3 significant figures the value of:

(a) $\sin\theta$, given that $\cos\theta = 0.271\,36$

(b) $\cos\theta$, given that $\tan\theta = 1.0123$

(c) $\tan\theta$, given that $\sin\theta = 0.8021$

(d) $\cos\theta$, given that $\sin\theta = 0.741\,59$

(e) $\sin\theta$, given that $\tan\theta = 0.982\,81$

(f) $\tan\theta$, given that $\cos\theta = 0.772\,16$

(g) $\sin\theta$, given that $\cos\theta = 0.210\,12$

(h) $\cos\theta$, given that $\tan\theta = 1.9527$

(i) $\tan\theta$, given that $\sin\theta = 0.334\,51$

(j) $\sin\theta$, given that $\tan\theta = 1.142\,41$

7.5 Using trigonometry to solve problems in two dimensions

The definitions of sine, cosine and tangent for acute angles can be used to help solve problems in two dimensions. So far our definitions of sine, cosine and tangent are limited to acute angles which lie in right-angled triangles, so you can only use them to solve problems in which the triangle contains a right angle.

Example 13

Find to 3 significant figures the length of the side AB in the triangle ABC.

$$\frac{AB}{2.871} = \tan 42.1^\circ$$

$$AB = 2.871 \times \tan 42.1^\circ$$

$$= 2.871 \times 0.903\,569\ldots$$

$$= 2.594\,14\ldots$$

$$AB = 2.59\,\text{cm} \ (3\,\text{s.f.})$$

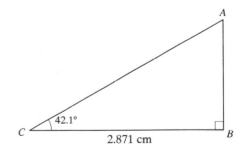

Example 14

Find to 3 significant figures the length of the side AB in the triangle ABC.

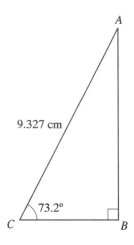

$$\frac{AB}{9.327} = \sin 73.2°$$

$$AB = 9.327 \times \sin 73.2°$$

$$= 9.327 \times 0.957\,319\,..$$

$$= 8.928\,918\,9\,..$$

$$AB = 8.93 \,\text{cm} \; (3 \text{ s.f.})$$

Example 15

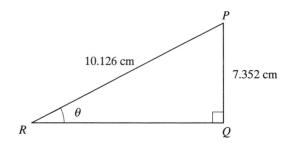

Find in degrees to 1 decimal place the size of the angle θ.

$$\sin \theta = \frac{7.352}{10.126}$$

$$= 0.726\,051\,7\,..$$

$$\theta = \arcsin 0.726\,051\,7\,..$$

$$= 46.556\,41\,..\,°$$

$$\theta = 46.6° \; (1 \text{ d.p.})$$

Example 16

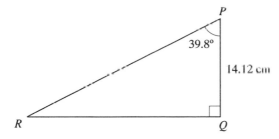

Calculate to 3 significant figures the length of the side PR.

$$\frac{14.12}{PR} = \cos 39.8°$$

$$\frac{PR}{14.12} = \frac{1}{\cos 39.8°}$$

$$PR = \frac{14.12}{\cos 39.8°}$$

$$= \frac{14.12}{0.768\,283\ldots}$$

$$= 18.378\,63\ldots$$

$$PR = 18.4\,\text{cm} \ (3\,\text{s.f.})$$

Example 17

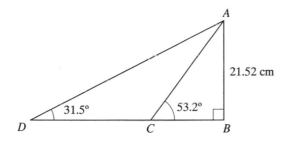

Calculate to 3 significant figures the length of CD.

$$\angle BAC = 90° - 53.2° = 36.8°$$

So:

$$\frac{BC}{21.52} = \tan 36.8°$$

$$BC = 21.52 \times \tan 36.8°$$

$$= 21.52 \times 0.748\,095$$

$$BC = 16.099\,01\ldots \text{cm}$$

$$= 16.10\,\text{cm} \ (4\,\text{s.f.})$$

$$\angle BAD = 90° - 31.5° = 58.5°$$

So:

$$\frac{BD}{21.52} = \tan 58.5°$$

$$BD = 21.52 \times \tan 58.5°$$

$$= 21.52 \times 1.631\,851\,6..$$

$$BD = 35.117\,448.. \text{ cm}$$

$$= 35.12\,\text{cm}(4\,\text{s.f.})$$

$$CD = BD - BC = 35.12 - 16.10$$

$$CD = 19.0\,\text{cm}(3\,\text{s.f.})$$

Notice that in questions such as this each length must be stored to at least 4 significant figures if you want to give the final answer to 3 significant figures. If you approximate too soon you will usually get an inaccurate final answer.

Exercise 7C

1 Find the sizes of the angles and the lengths of the sides marked.
Give the angles in degrees to 1 decimal place and the sides in cm
to 3 significant figures.

(a)

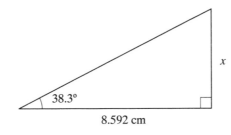

(b)

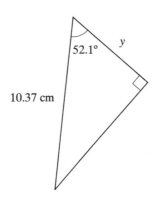

(c)

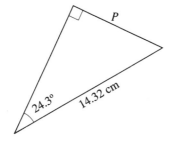

(d)

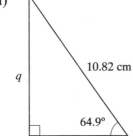

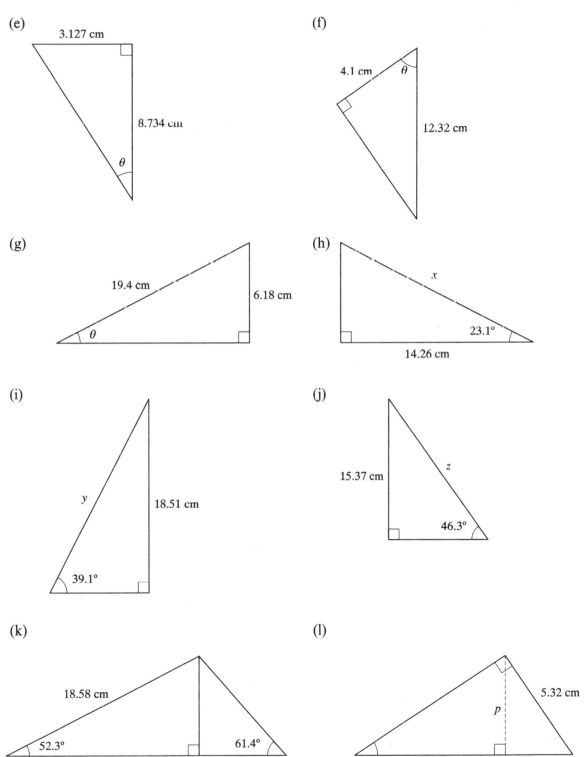

(e)

3.127 cm

8.734 cm

θ

(f)

4.1 cm

θ

12.32 cm

(g)

19.4 cm

6.18 cm

θ

(h)

x

23.1°

14.26 cm

(i)

y

18.51 cm

39.1°

(j)

15.37 cm

z

46.3°

(k)

18.58 cm

52.3°

61.4°

y

(l)

5.32 cm

p

17.59 cm

(m)

(n)

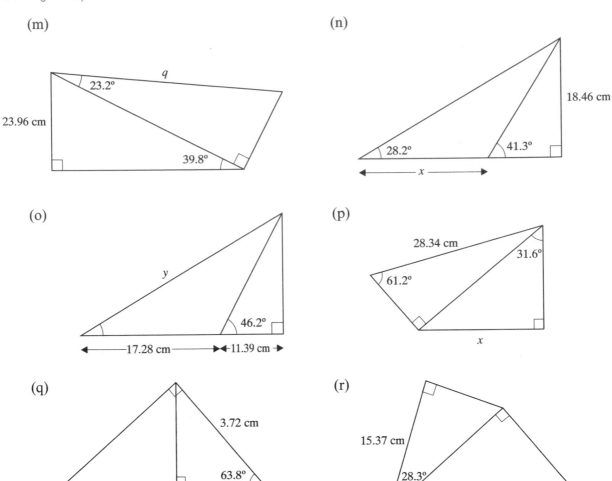

(o)

(p)

(q)

(r)

2

The figure shows triangle ABC in which $AC = 15.6\,\text{cm}$, $\angle ABC = 90°$ and $\angle BCA = 70.3°$. The perpendicular from B to AC meets AC in N.

Calculate to 3 significant figures:

(a) the length, in cm, of *BN*

(b) the area, in cm², of the triangle *ABC*.

3

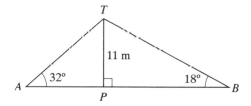

The vertical flagpole *PT* is of height 11 m and stands on level ground at *P*. The points *A* and *B* are on the same horizontal level as *P* and ∠*TAP* = 32°, ∠*TBP* = 18°.

(a) Calculate, in m to one decimal place, the length *AP*.

(b) Calculate, in m to one decimal place, the length *BT*.

[L]

4

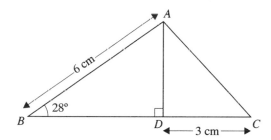

In the figure *D* is the point on *BC* such that ∠*ADB* is a right angle. Given that *AB* = 6 cm, *DC* = 3 cm and ∠*ABD* = 28°, calculate:

(a) *AD* (b) *BD* (c) ∠*ACD*.   [L]

5

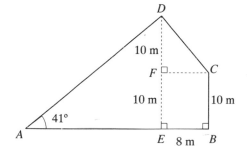

In the figure, *ABCD* represents the cross-section through a building. The perpendicular from *D* meets *AB* in *E*. *CF* is perpendicular to *DE*. Angle *DAE* = 41°.

$DF = FE = CB = 10$ m and $EB = 8$ m.

Calculate to 3 significant figures:

(a) angle DCF

(b) the length, in metres, of DC

(c) the length, in metres, of AB

(d) the area, in square metres, of $ABCD$. [L]

6

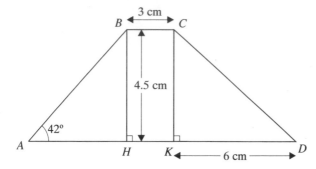

In the figure $ABCD$ is a trapezium with BC parallel to AD. The lines BH and CK are each perpendicular to AD and are 4.5 cm in length. Given that $BC = 3$ cm, $KD = 6$ cm and $\angle BAH = 42°$, calculate:

(a) CD

(b) AB

(c) AD

(d) the perimeter of the trapezium $ABCD$

(e) the area of the trapezium $ABCD$. [L]

7

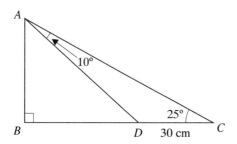

In the figure the triangle ABC is right-angled at B and the point D, on BC, is such that $DC = 30$ cm. Given that $\angle ACD = 25°$ and $\angle DAC = 10°$, calculate the area of triangle ABC, giving your answer in square centimetres to three significant figures.

[L]

8

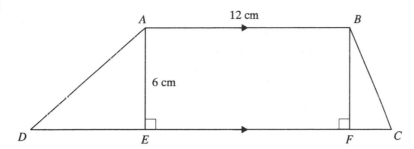

The figure shows the trapezium $ABCD$ in which AB is parallel to DC, $AB = 12$ cm, $\angle ABC = 117°$ and $\angle ADC = 46°$. The perpendicular from A to DC meets DC in E and $AE = 6$ cm. The perpendicular from B to DC meets DC in F. Calculate, in centimetres, giving your answers to 3 significant figures:

(a) the length of DE

(b) the length of AD

(c) the length of FC.

Find also, in square centimetres, giving your answer to 3 significant figures, the area of the trapezium. [L]

9 The point O is at the top of a vertical cliff. From a boat which is at a point A, due south of O, the angle of elevation of O is $28°$. The boat now sails 50 metres due north to a point B from which the angle of elevation of O is $70°$. Calculate, in metres to 3 significant figures:

(a) the height of the cliff

(b) the distance of B from the foot of the cliff. [L]

10 A ship sails from port A to port B, a distance of 5 km, on a bearing of $036°$.

(a) Calculate, in km to 2 decimal places, the distance by which B is

(i) east of A

(ii) north of A.

The ship then sails to a point C, a further distance of 8 km, on a bearing of $138°$.

(b) Calculate, in km to 2 decimals places, the distance by which C is:

(i) east of A

(ii) south of A.

(c) Calculate:

(i) the bearing, to the nearest degree, of A from C

(ii) the distance, in km to 2 decimal places, of A from C.

[L]

11

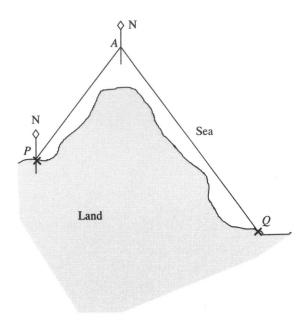

A ship leaves a Port P and travels a distance of 12 km on a bearing of 037° to a buoy A. The ship then changes course and travels 25 km on a bearing of 140° to the port Q. Calculate, in km, to 2 decimal places:

(a) how far A is to the north and to the east of P

(b) how far Q is to the south and to the east of A

(c) the shortest distance from P to Q.

Calculate, to the nearest degree,

(d) the bearing of Q from P.   [L]

7.6 Some special results

Look at the ratio $\dfrac{\sin\theta}{\cos\theta}$.

$$\frac{\sin\theta}{\cos\theta} = \frac{\dfrac{BC}{AC}}{\dfrac{AB}{AC}} = \frac{BC}{AC} \times \frac{AC}{AB} = \frac{BC}{AB} = \tan\theta$$

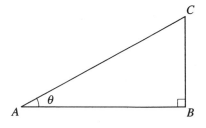

■ **So for any given angle,** $\dfrac{\sin\theta}{\cos\theta} = \tan\theta$

This is a very useful result and should be remembered.

$$\angle ACB = 90° - \theta \text{ and } \sin\angle ACB = \sin(90° - \theta) = \frac{AB}{AC}$$

However, $\dfrac{AB}{AC} = \cos\theta$

■ **So for any given acute angle θ,**

$$\cos\theta = \sin(90° - \theta)$$

Similarly, $\sin\theta = \cos(90° - \theta)$

Memorise these.

Consider an equilateral triangle ABC of side 2.

Obviously, each of its angles is of size 60°.

If you drop a perpendicular from A to X, the mid-point of BC, then $\angle XAB = 30°$ and $CX = XB = 1$.

Use Pythagoras' theorem in the right-angled triangle AXB:

$$AX^2 = 2^2 - 1^2 = 4 - 1 = 3$$

So $AX = \sqrt{3}$ units

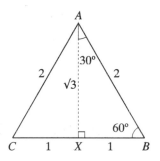

Thus in the triangle AXB:

$$\sin 30° = \tfrac{1}{2}$$

$$\cos 30° = \frac{\sqrt{3}}{2}$$

and $$\tan 30° = \frac{1}{\sqrt{3}} \text{ or } \frac{\sqrt{3}}{3}$$

Also, $$\sin 60° = \frac{\sqrt{3}}{2}$$

$$\cos 60° = \tfrac{1}{2}$$

and $$\tan 60° = \sqrt{3}$$

Now consider an isosceles, right-angled triangle PQR in which $PQ = QR = 1$. As the triangle is isosceles $\angle QPR = \angle QRP = 45°$. Pythagoras' theorem gives:

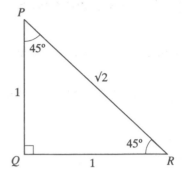

$$PR^2 = 1^2 + 1^2 = 2$$
$$PR = \sqrt{2}$$

So: $\qquad \sin 45° = \cos 45° = \dfrac{1}{\sqrt{2}} \text{ or } \dfrac{\sqrt{2}}{2}$

and $\qquad \tan 45° = \dfrac{1}{1} = 1$

■ **The angles $30°$, $45°$ and $60°$ occur frequently in trigonometry. Memorise their exact trigonometric ratios.**

	30°	**45°**	**60°**
sin	$\frac{1}{2}$	$\frac{1}{\sqrt{2}}$	$\frac{\sqrt{3}}{2}$
cos	$\frac{\sqrt{3}}{2}$	$\frac{1}{\sqrt{2}}$	$\frac{1}{2}$
tan	$\frac{1}{\sqrt{3}}$	1	$\sqrt{3}$

7.7 The three basic trigonometric functions for *any* angle

The theory of trigonometry which applies to acute angles can be extended to angles of any size. Think of an arm, OA, that is fixed at O and turns in an anticlockwise direction, starting along the positive x-axis. As it turns, it makes an angle with the positive x-axis.

The quadrant between the positive x-axis and the positive y-axis is called the **first quadrant**. In this quadrant any angle θ is acute. So in this diagram θ is an angle such that $0 < \theta < 90°$

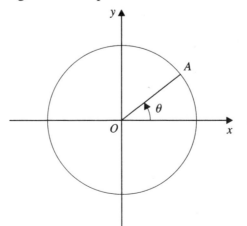

(or $0 < \theta < \dfrac{\pi^c}{2}$).

The three trigonometric ratios for positive, acute angles have already been defined (on p. 127).

The angle θ is in the **second quadrant** when the arm OA lies between the negative x-axis and the positive y-axis. Here the angle is obtuse – it lies between 90° and 180° (or $\dfrac{\pi^c}{2}$ and π^c).

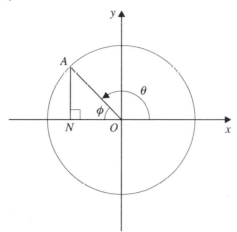

To find the trigonometric ratios of angles in the second quadrant drop a perpendicular from A to the x-axis at N and consider the acute angle ϕ which OA makes with the x-axis. In the triangle ANO, the length of the arm AO is defined to be a positive quantity. AN is upwards in the positive y direction so its length is a positive quantity. ON is in the negative x direction and so is negative. We define $\sin\theta$, $\cos\theta$ and $\tan\theta$ to be equal in size to $\sin\phi$, $\cos\phi$ and $\tan\phi$ but with the appropriate sign.

Thus
$$\sin\phi = \frac{AN}{AO} = \frac{+\text{ive}}{+\text{ive}} = +\text{ive}$$

$$\cos\phi = \frac{NO}{AO} = \frac{-\text{ive}}{+\text{ive}} = -\text{ive}$$

$$\tan\phi = \frac{AN}{NO} = \frac{+\text{ive}}{-\text{ive}} = -\text{ive}$$

So in the second quadrant:

$$\sin\theta = +\sin\phi$$
$$\cos\theta = -\cos\phi$$
$$\tan\theta = -\tan\phi$$

When the position of the arm OA is such that $180° < \theta < 270°$, it is said to be in the **third quadrant**. As before, to find the trigonometric ratios in the third quadrant, drop a perpendicular from A to N on the x-axis and consider the acute angle ϕ lying in the right-angled triangle AON.

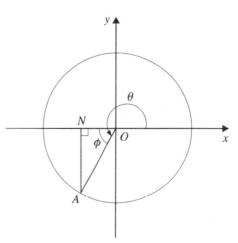

AO is again defined positive.

ON is negative.

AN is negative.

So in the third quadrant:

$$\sin\phi = \frac{AN}{AO} = \frac{-\text{ive}}{+\text{ive}} = -\text{ive}$$

$$\cos\phi = \frac{NO}{AO} = \frac{-\text{ive}}{+\text{ive}} = -\text{ive}$$

$$\tan\phi = \frac{AN}{NO} = \frac{-\text{ive}}{-\text{ive}} = +\text{ive}$$

Again, we define $\sin\theta$, $\cos\theta$ and $\tan\theta$ in the third quadrant to be $\sin\phi$, $\cos\phi$ and $\tan\phi$ with the appropriate signs.

So in the third quadrant:

$$\sin\theta = -\sin\phi$$
$$\cos\theta = -\cos\phi$$
$$\tan\theta = +\tan\phi$$

In the **fourth quadrant**, when $270° < \theta < 360°$, the perpendicular *AN* is again dropped to the *x*-axis so that the acute angle ϕ can be considered in the triangle *ANO*.

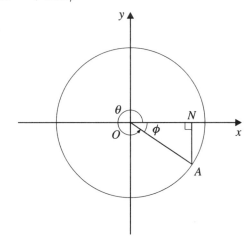

In this quadrant:

AO is positive.

ON is positive.

AN is negative.

In the fourth quadrant:

$$\sin\phi = \frac{AN}{AO} = \frac{-\text{ive}}{+\text{ive}} = -\text{ive}$$

$$\cos\phi = \frac{NO}{AO} = \frac{+\text{ive}}{+\text{ive}} = +\text{ive}$$

$$\tan\phi = \frac{AN}{NO} = \frac{-\text{ive}}{+\text{ive}} = -\text{ive}$$

$\sin \theta$, $\cos \theta$ and $\tan \theta$ are defined as before to be $\sin \phi$, $\cos \phi$ and $\tan \phi$ with the appropriate signs.

That is, in the fourth quadrant:

$$\sin \theta = - \sin \phi$$
$$\cos \theta = + \cos \phi$$
$$\tan \theta = - \tan \phi$$

Angles can lie outside the range 0–360° but they always lie in one of the four quadrants and so the same results apply.

The angle 410° lies in the first quadrant.

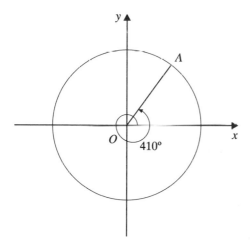

If the arm OA moves in a clockwise direction, then the angle through which it turns is defined to be negative.

The angle −140° lies in the third quadrant.

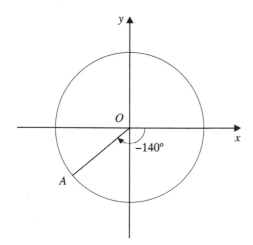

The angle $-250°$ lies in the second quadrant.

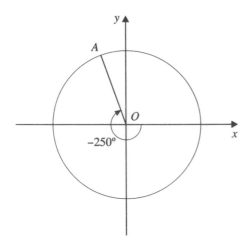

Remember, always drop a perpendicular from A to N on the x-axis (*never* to the y-axis), consider the acute angle in the triangle ANO and give it an appropriate sign.

■ **The easiest way to remember which trigonometric ratios are positive and which negative in each quadrant is to remember which ratios are positive and remember that the others are negative. In the first quadrant all the ratios are positive, in the second quadrant the sine is positive, in the third quadrant the tangent is positive and in the fourth quadrant cosine is positive. These results can be shown diagrammatically.**

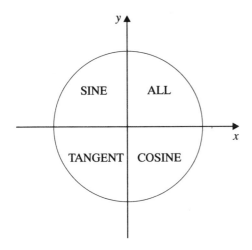

A good way to remember these is by the mnemonic 'positively <u>A</u>ll <u>S</u>ilver <u>T</u>ea <u>C</u>ups'.

Example 18

Write $\cos 127°$ as a trigonometric ratio of an acute angle.

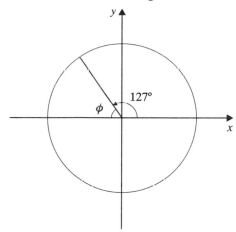

$127°$ lies in the second quadrant. So consider

$$\phi = 180° - 127° = 53°$$

In the second quadrant cosine is negative, so:

$$\cos 127° = -\cos 53°$$

Example 19

Write $\sin(-120°)$ as the trigonometric ratio of an acute angle.

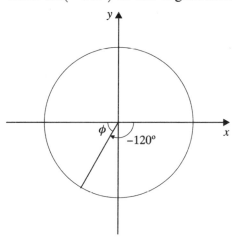

$-120°$ lies in the third quadrant, so consider

$$\phi = 180° - 120° = 60°$$

In the third quadrant sine is negative so

$$\sin(-120°) = -\sin 60°$$

Example 20

Write $\sin 380°$ as the trigonometric ratio of an acute angle.

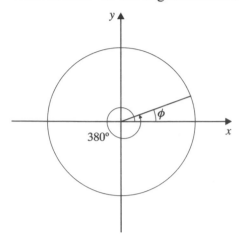

$380°$ lies in the first quadrant. So consider

$$\phi = 380° - 360° = 20°$$

Sine is positive in the first quadrant, so

$$\sin 380° = +\sin 20°$$

Example 21

Find $\sin 315°$, leaving your answer in terms of surds.

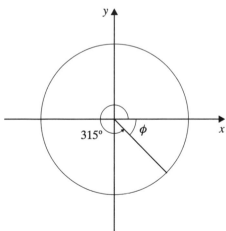

$315°$ lies in the fourth quadrant, so

$$\phi = 360° - 315° = 45°$$

In the fourth quadrant sine is negative, so

$$\sin 315° = -\sin 45° = -\frac{1}{\sqrt{2}}$$

Example 22

Solve the equation $\cos x = 0.3941$ for $0 < x \leqslant 360°$

$$\cos x = 0.3941$$

So: $$x = \arccos 0.3941$$
$$= 66.8° \text{ from a calculator}$$

But cosine is positive in the first and fourth quadrants. So a second solution lies in the fourth quadrant:

$$x = 360° - 66.8° = 293.2°$$

So the solutions are $x = 66.8°$ and $293.2°$.

Example 23

Solve the equation:

$$\sin 2x = \tfrac{1}{2} \text{ for } 0 < x \leqslant 2\pi$$

$$\sin 2x = \tfrac{1}{2}$$

So: $$2x = \arcsin \tfrac{1}{2}$$

and: $$2x = 30° = \frac{\pi}{6}$$

But sine is positive in the first *and* second quadrants. So a second solution is:

$$2x = \pi - \frac{\pi}{6} = \frac{5\pi}{6}$$

Another solution in the first quadrant is:

$$2x = 2\pi + \frac{\pi}{6} = \frac{13\pi}{6}$$

and yet another in the second quadrant is:

$$2x = 2\pi + \frac{5\pi}{6} = \frac{17\pi}{6}$$

Dividing by 2 gives: $x = \dfrac{\pi}{12}, \dfrac{5\pi}{12}, \dfrac{13\pi}{12}$ and $\dfrac{17\pi}{12}$.

Notice that the equation $\sin x = \tfrac{1}{2}$ has two solutions, namely $\dfrac{\pi}{6}$ and $\dfrac{5\pi}{6}$ in the interval $(0, 2\pi)$.

Notice too that $\sin 2x = \tfrac{1}{2}$ has four solutions in the same interval. Similarly, $\sin 3x = \tfrac{1}{2}$ has six solutions: you should check this statement for yourself.

Exercise 7D

1   Write each of the following as trigonometric ratios of positive acute angles:

(a)	$\sin 260°$	(b)	$\cos 140°$	(c)	$\tan 185°$
(d)	$\tan 355°$	(e)	$\cos 137°$	(f)	$\sin 414°$
(g)	$\sin(-194°)$	(h)	$\cos(-336°)$	(i)	$\tan 396°$
(j)	$\tan 148°$	(k)	$\sin(-443)°$	(l)	$\cos 248°$
(m)	$\sin 331°$	(n)	$\cos 293°$	(o)	$\tan(-96°)$
(p)	$\tan(216°)$	(q)	$\sin 339°$	(r)	$\cos 481°$
(s)	$\tan 127°$	(t)	$\sin(-198°)$		

2   Write down the values of the following leaving your answers in terms of surds where appropriate:

(a)	$\sin 120°$	(b)	$\cos 150°$	(c)	$\tan 225°$
(d)	$\cos 300°$	(e)	$\sin(-30°)$	(f)	$\cos(-120°)$
(g)	$\sin 240°$	(h)	$\sin 420°$	(i)	$\cos 315°$
(j)	$\sin(-135°)$	(k)	$\tan(-150°)$	(l)	$\tan 150°$
(m)	$\sin 330°$	(n)	$\cos 210°$	(o)	$\tan 240°$
(p)	$\tan 390°$	(q)	$\cos(-300°)$	(r)	$\cos(-420°)$
(s)	$\sin 405°$	(t)	$\tan 150°$		

3   Solve the following for $0 < x \leqslant 360°$ giving x to 1 decimal place.

(a)	$\sin x = 0.332$	(b)	$\tan x = -1.102$
(c)	$\cos x = -0.7486$	(d)	$\tan x = 0.3842$
(e)	$\cos(x + 20°) = 0.7615$	(f)	$\tan(x - 33°) = 0.9451$
(g)	$\sin 2x = 0.4751$	(h)	$\cos 3x = 0.5196$
(i)	$\sin(2x - 20°) = 0.6348$	(j)	$\sin(70° - x) = 0.3313$
(k)	$\cos(\frac{1}{2}x - 39.6°) = 0.9144$	(l)	$\tan(3x - 40°) = 0.61$
(m)	$\sin x - (3 \sin x - 1) = 0$	(n)	$\cos x(4 \sin x - 3) = 0$
(o)	$8 \sin^2 x - 6 \sin x + 1 = 0$	(p)	$\tan^2 x + 2 \tan x + 1 = 0$

4   Find the values of x in the interval $0 < x < \pi$ which satisfy the equation

$$\cos 2x = -0.4$$

giving your answers in radians to 2 decimal places.   [L]

5   Given that $0 < \theta < 180$, find θ when:
(a) $\tan \theta° = -1$
(b) $\sin 2\theta° = -\frac{1}{2}$ [L]

6 Find the values of θ in the interval $0 \leqslant \theta < 2\pi$ which satisfy
   (a) $\cos\theta = -0.8$, giving your answers to 3 significant figures
   (b) $\sin(\theta - \frac{1}{3}\pi) = 0.5$, giving your answers as multiples of π.

[L]

7.8 Drawing the graphs of $\sin x$, $\cos x$ and $\tan x$

Section 7.7 shows how to find the sine, cosine or tangent of any angle. This section shows you how to draw the graphs of these functions.

Graphing $\sin x$

Here is a table of values for the function $y = \sin x$ when $0 \leqslant x \leqslant 360°$. Each value of y is given to two decimal places:

x	0	30°	45°	60°	90°	120°	135°	150°	180°	210°	225°	240°	270°	300°	315°	330°	360°
y	0	0.5	0.71	0.87	1	0.87	0.71	0.5	0	−0.5	−0.71	−0.87	−1	−0.87	−0.71	−0.5	0

If you plot these figures on a graph, it looks like this.

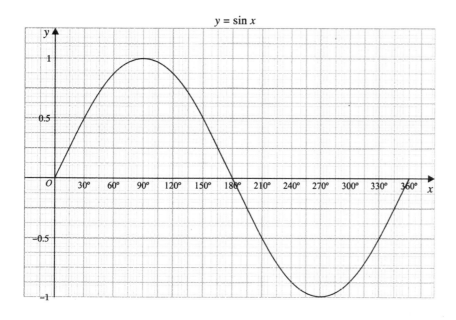

$y = \sin x$

The curve will repeat itself again and again for values of x above $360°$ and below 0. The function has a maximum value of $+1$ and a minimum value of -1. The maximum value occurs at $90°$, $450°$, $810°$, and so on; in other words at $90° \pm 360n°$, where n is an integer (or at $\frac{\pi^c}{2} \pm 2n\pi^c$). The minimum value occurs at $270°$, $630°$, $990°$, and so on; in other words at $270° \pm 360n°$ (or at $\frac{3\pi^c}{2} \pm 2n\pi^c$).

The other main feature of the curve is that it cuts the x-axis at 0, $180°$, $360°$, $540°$, $720°$, and so on; in other words every $180n°$ (or $n\pi^c$).

As the curve starts at O and continues to $360°$ before repeating itself, we say that the sine curve has a **period** of $360°$ (or $2\pi^c$).

Graphing cos x

Here is a table of values for the function $y = \cos x$ when $0 \leqslant x \leqslant 360°$. Again each value of y is given to 2 decimal places:

x	0	30°	45°	60°	90°	120°	135°	150°	180°	210°	225°	240°	270°	300°	315°	330°	360°
y	1	0.87	0.71	0.5	0	−0.5	−0.71	−0.87	−1	−0.87	−0.71	−0.5	0	0.5	0.71	0.87	1

The curve looks like this:

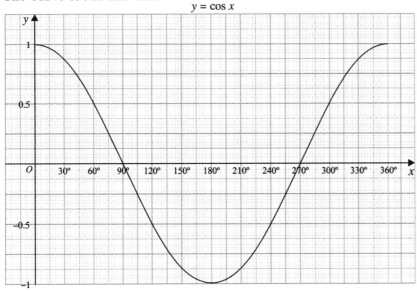

Again the main feature of the curve is that it is periodic (or cyclic). It has a period of $360°$. Its maximum is again 1 and its minimum -1. But this time the maximum value occurs at 0, $360°$, $720°$, $1080°$, and so on; in other words at $360n°$ (or $2n\pi^c$). The minimum value occurs at 180°, 540°, 900°, and so on; in other words at $180° \pm 360n°$ (or at $\pi^c \pm 2n\pi^c$). The curve cuts the x-axis at 90°, 270°, 450°, 630°, and so on; in other words at $\pm(2n + 1)90°$ or $\pm(2n + 1)\frac{\pi^c}{2}$.

Graphing tan x

Here is a table of values for $y = \tan x$ when $0 \leqslant x \leqslant 360°$:

x	0	30°	45°	60°	90°	120°	135°	150°	180°	210°	225°	240°	270°	300°	315°	330°	360°
y	0	0.58	1	1.73	∞	−1.73	−1	−0.58	0	0.58	1	1.73	∞	−1.73	−1	−0.58	0

(The symbol ∞ means 'infinity')

The curve looks like this:

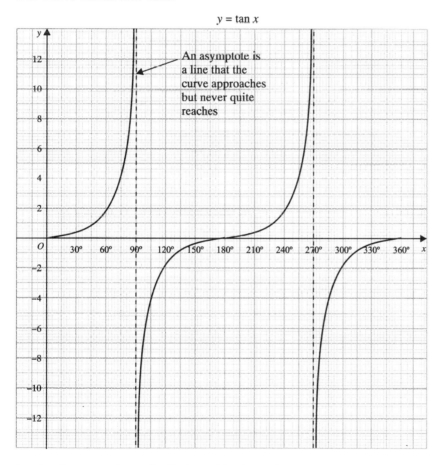

Again the curve is periodic but this time the period is 180° (π^c). The curve cuts the x-axis at 0, 180°, 360°, 540°, 720°, and so on; in other words at $180n°$ (or $n\pi^c$). It has **asymptotes** (lines that the curve approaches but never actually reaches) at 90°, 270°, 450°, and so on; in other words at $\pm(2n + 1)90°$ (or $\pm(2n + 1)\frac{\pi^c}{2}$). The graph of $y = \tan x$ has no maximum or minimum.

You must memorise the main features of the curves $y = \sin x$, $y = \cos x$ and $y = \tan x$ – their shape, their maximum and minimum values and where they occur, the points at which they cut the x-axis, and the position of the asymptotes. For an advanced course in mathematics you must be able to sketch these three curves from memory.

7.9 Simple transformations of the functions sin x, cos x and tan x

This section shows the effect of simple transformations on the graphs of $y = \sin x$, $y = \cos x$ and $y = \tan x$.

Graphing sin nx, cos nx and tan nx

What happens to functions such as $y = \sin x$ when x is multiplied by a number to give $y = \sin 2x$, $y = \sin 3x$ and so on? Here is the table of values of $y = \sin 2x$:

x	0	30°	45°	60°	90°	120°	135°	150°	180°	210°	225°	240°	270°	300°	315°	330°	360°
y	0	0.87	1	0.87	0	−0.87	−1	−0.87	0	0.87	1	0.87	0	−0.87	−1	−0.87	0

The curve looks like this:

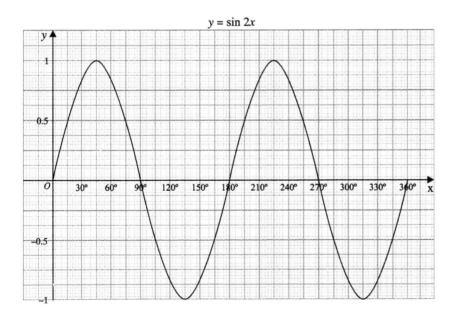

The curve of $y = \sin 2x$ is the same shape as the curve of $y = \sin x$ but the period is 180°. In other words *two* sine curves fit into the range 0–360°. This is generally true of this type of transformation.

The sketch of $y = \cos\frac{1}{2}x$ looks like this:

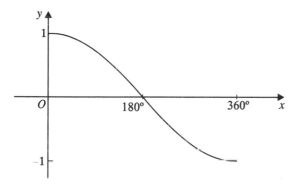

The curve of $y = \cos\frac{1}{2}x$ is the same shape as the curve of $y = \cos x$ but only *half* the graph appears in the range 0–360°.

■ **In general the curve of $y = \sin nx$, $y = \cos nx$ or $y = \tan nx$ where n is a positive real number ($n \in \mathbb{R}^{+}$) is the same shape as the curve of $y = \sin x$, $y = \cos x$ or $y = \tan x$, but n of each curve fit into the range 0–360°.**

Example 24

Sketch the graph of $y = \tan 3x$ in the range $0 \leqslant x \leqslant 360°$.

$y = \tan 3x$ tells us that three curves with the same shape as $y = \tan x$ fit into the range $0 \leqslant x \leqslant 360°$. The curve looks like this:

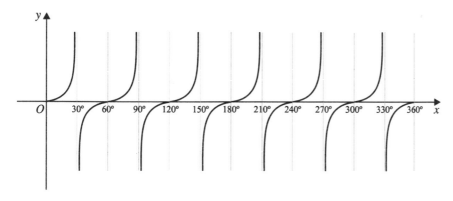

Example 25

Sketch the curve of $y = \cos\frac{1}{3}x$ in the range $0 \leqslant x \leqslant 360°$.

The shape of $y = \cos\frac{1}{3}x$ is the same as that of $y = \cos x$ but only one-third of the curve fits in the range $0 \leqslant x \leqslant 360°$. So the sketch of $y = \cos\frac{1}{3}x$ looks like this:

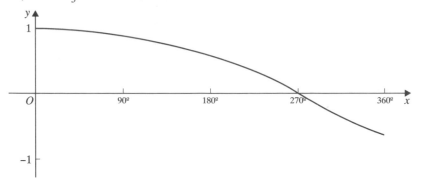

Graphing $-\sin x$, $-\cos x$ and $-\tan x$

Transformations such as $y = -\sin x$ are easy to deal with as for any function the curve of $y = -f(x)$ is a reflection of the curve of $y = f(x)$ in the x-axis.

Here is the curve of $y = -\sin x$ for $0 \leqslant x \leqslant 360°$:

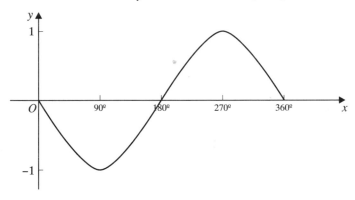

Graphing $\sin(-x)$, $\cos(-x)$ and $\tan(-x)$

It is harder to graph the transformation that occurs when the minus sign is *inside* the function: for example, $y = \sin(-x)$. From the work done on p. 147 on the trigonometric ratios of negative angles:

$$\sin(-x) = -\sin x$$

The graph of $y = \sin(-x)$ is the same as that shown above for $y = -\sin x$.

$$\cos(-x) = +\cos x$$
$$\tan(-x) = -\tan x$$

You can also check these from the graphs of the functions.

To draw the curve of $y = \cos(-x)$ rewrite the function as $y = \cos x$. The sketch of the curve is the same as that of $y = \cos x$ on p. 152.

Example 26
Sketch the curve of $y = \tan(-2x)$.

$y = \tan(-2x)$ can be written $y = -\tan 2x$.

The graph of $y = \tan 2x$ has the same shape as the graph of $y = \tan x$, but two of these curves fit into $0 \leqslant x \leqslant 360°$. The graph of $y = -\tan 2x$ is a reflection of the graph of $y = \tan 2x$ in the x-axis. Here is the sketch:

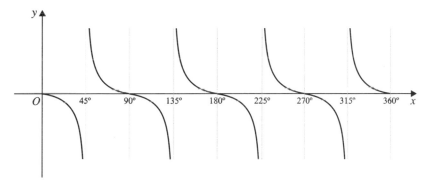

Graphing $n \sin x$, $n \cos x$ and $n \tan x$

The curve of $y = 3f(x)$ represents a stretch of the curve $y = f(x)$ from the x-axis of scale factor 3. The transformation $y = 3\sin x$ is a stretch of the curve $y = \sin x$ from the x-axis of scale factor 3. The transformation $y = 5\tan x$ is a stretch of the curve $y = \tan x$ from the x-axis of scale factor 5. For example, here is the curve of $y = 3\cos x$:

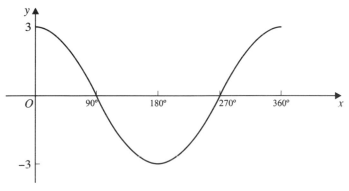

Graphing sin(x + n), cos(x + n) and tan(x + n)

The final transformation we shall deal with is one of the form $y = f(x + n)$ where n is a constant. Here is a table of values for the function $y = \sin(x + 30°)$:

x	0	30°	45°	60°	90°	120°	135°	150°	180°	210°	225°	240°	270°	300°	315°	330°	360°
y	0.5	0.87	0.97	1	0.87	0.5	0.26	0	−0.5	−0.87	−0.97	−1	−0.87	−0.5	−0.26	0	0.5

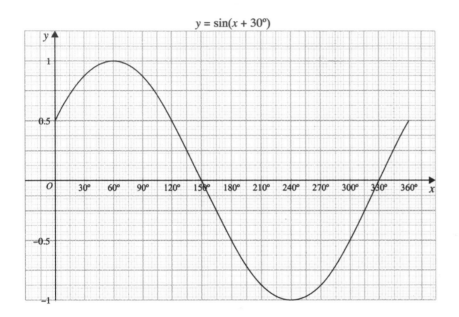

$y = \sin(x + 30°)$

The curve is the same shape as that of $y = \sin x$ but it has been shifted 30° *to the left*.

- **Generally, the graph of $y = f(x + n)$ is the same as that of $y = f(x)$ but translated n to the left. Similarly, the graph of $y = f(x − n)$ is the same as that of $y = f(x)$ but translated n to the right.**

Remember that when a *positive* number is added the graph is translated that amount in the *negative x* direction. When a *negative* number is added the graph is translated that amount in the *positive x* direction.

Example 27

Sketch the curve of $y = 1 + 2\sin(x - 90°)$ for $0 \leqslant x \leqslant 360°$.

The '$-90°$' means that the sine graph is translated $90°$ to the right:

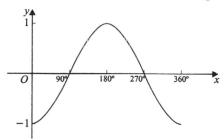

The '2' means that it is then stretched by a factor 2 from the x-axis:

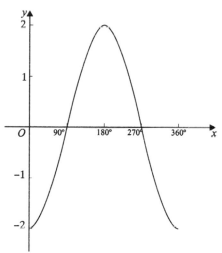

Finally 1 is added to all the y values:

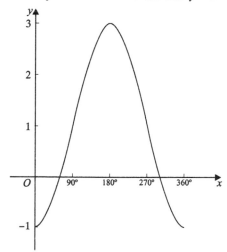

Exercise 7E

1   Plot the graph of $y = 1 + \cos 2x$ for $0 \leqslant x \leqslant 360°$. From your graph find approximate solutions to:

(a) $1 + \cos 2x = 1.7$ (b) $1 + \cos 2x = 0.3$

(c) $\cos 2x = 0.3$.

2   Plot the graph of $y = \cos \frac{1}{2}x$ for $0 \leqslant x \leqslant 360°$. Hence find approximate solutions to:

(a) $\cos \frac{1}{2}x = 0.7$ (b) $\cos \frac{1}{2}x = -0.3$.

3   Plot on the same axes the graphs of $y = \sin x$ and $y = 0.5 + \cos x$ for $0 \leqslant x \leqslant 360°$.

Use your graph to find approximate solutions of:

(a) $\sin x = 0.5 + \cos x$ (b) $0.5 + \cos x = 1.1$

(c) $\cos x = 0.3$

4   Given that:

$$f(x) = \tan x° + \frac{2}{\tan x°}$$

copy and complete the following table at the top of a piece of graph paper.

x	35	40	45	50	55	60	65	70
$f(x)$	3.56	3.22	3			2.89	3.08	3.48

Draw an accurate graph of $y = f(x)$ for $35 \leqslant x \leqslant 70$.

Use your graph to estimate the set of values of x, within the above range, for which

$$f(x) \leqslant 3.3$$ [L]

5   (a) Copy and complete the following table for which

$$f(x) \equiv \cos^2 x° - \sin x°, \ 0 \leqslant x \leqslant 180$$

giving your answers to 2 decimal places where appropriate.

x	0	30	60	90	120	150	180
$f(x)$	1	0.25					

(b) On graph paper and using a scale of 2 cm to represent 30° on the x-axis and 4 cm to represent 1 unit on the y-axis, draw the graph of $y = f(x)$.

(c) From your graph estimate the solutions of the equation

$$\cos^2 x° - \sin x° = 0, \ 0 \leqslant x \leqslant 180.$$ [L]

6 Sketch the graphs of the following for $0 \leqslant x \leqslant 360°$:

(a) $y = -\cos x$ (b) $y = 2\sin x$

(c) $y = \tan(-x)$ (d) $y = \tan(x + 30°)$

(e) $y = -3\sin x$ (f) $y = 1 + \tan x$

(g) $y = 1 - 2\cos x$ (h) $y = 1 + \cos(-2x)$

(i) $y - 2\cos(-2x)$ (j) $y - 2\sin(x + 60°)$

7 Sketch the graphs of the following for $-180° \leqslant x \leqslant 180°$:

(a) $y = \sin(x - 30°)$ (b) $y - \cos(x + 30°)$

(c) $y = \tan(x + 60°)$ (d) $y = 1 + 2\cos x$

(e) $y = \cos(60° - x)$ (f) $y = \sin(30° - x)$

(g) $y = 2\tan(90° - x)$ (h) $y = 1 + 3\cos(90° - x)$

(i) $y - 1 - 25\sin\frac{1}{3}x$ (j) $y = 1 - \frac{1}{2}\sin 2x$

8 Using the same scales and axes, sketch, for $-\pi \leqslant x \leqslant \pi$, the graphs of:

(a) $y = \sin x$ (b) $y = \dfrac{2x}{\pi}$

where x is measured in radians.

On your sketch mark the two points P and Q other than the origin where your graphs intersect.

Give the coordinates of P and Q. [L]

9 (a) Sketch the graph of $y = \tan x$ for $0 \leqslant x \leqslant 3\pi$.

(b) Find the solutions, in the range $0 \leqslant x \leqslant 3\pi$, of the equation

$$\tan^2 x = 3$$ [L]

SUMMARY OF KEY POINTS

1 1 radian is the angle subtended at the centre of a circle by an arc with length that is equal to the radius of the circle.

2 The length of an arc of a circle, radius r, which subtends an angle θ^c at the centre of the circle is $r\theta$.

3 The area of a sector of a circle, radius r, bounded by an arc which subtends an angle θ^c at the centre of the circle is $\frac{1}{2}r^2\theta$.

4 In any right-angled triangle:

$$\sin\theta = \frac{\text{opposite}}{\text{hypotenuse}}$$

$$\cos\theta = \frac{\text{adjacent}}{\text{hypotenuse}}$$

$$\tan\theta = \frac{\text{opposite}}{\text{adjacent}}$$

5 If $\sin\theta = a$, then $\theta = \arcsin a$.
 If $\cos\theta = b$, then $\theta = \arccos b$.
 If $\tan\theta = c$, then $\theta = \arctan c$.

6 $\dfrac{\sin\theta}{\cos\theta} = \tan\theta$

7 $\cos(90° - \theta) = \sin\theta$
 $\sin(90° - \theta) = \cos\theta$

8 $\sin 30° = \frac{1}{2}$ $\qquad \cos 30° = \frac{\sqrt{3}}{2}$ $\qquad \tan 30° = \frac{1}{\sqrt{3}}$

9 $\sin 60° = \frac{\sqrt{3}}{2}$ $\qquad \cos 60° = \frac{1}{2}$ $\qquad \tan 60° = \sqrt{3}$

10 $\sin 45° = \frac{1}{\sqrt{2}}$ $\qquad \cos 45° = \frac{1}{\sqrt{2}}$ $\qquad \tan 45° = 1$

11 In the second quadrant only $\sin\theta$ is positive.

12 In the third quadrant only $\tan\theta$ is positive.

13 In the fourth quadrant only $\cos\theta$ is positive.

14 The graph of $y = \sin nx (n \in \mathbb{R}^+)$ is the same shape as that of $y = \sin x$, but n of the curves fit into the interval $0 \leqslant x \leqslant 360°$. The same is true for cosine and tangent functions.

15 The graph of $y = -\sin x$ is a reflection of $y = \sin x$ in the x-axis. The same is true for cosine and tangent functions.

16 To sketch the graph of $y = \sin(-x)$, $y = \cos(-x)$ and $y = \tan(-x)$ we use the fact that $\sin(-x) = -\sin x$, $\cos(-x) = +\cos x$ and $\tan(-x) = -\tan x$.

17 The graph of $y = a\sin x (a \in \mathbb{R}^+)$ is a stretch of the graph of $y = \sin x$, scale factor a, from the x-axis. The same is true for the cosine and tangent functions.

18 The graph of $y = \sin(x + n°)$ is the same shape as the graph of $y = \sin x$ but is translated $n°$ to the left. The same is true for cosine and tangent functions.

Differentiation

Differential calculus is the area of mathematics concerned with the rate at which things change. For example, the speed of a car is the rate at which the distance it travels changes with time.

This chapter starts by looking at the gradient of a straight line graph, which represents a rate of change.

8.1 The gradient of a graph

Gradient of a straight line graph

A straight line graph has a constant gradient or slope – one which is the same at any point on the line. The gradient can be found either from the equation of the line or by calculation from the coordinates of two points on the line. The gradient of the line between two points (x_1, y_1) and (x_2, y_2) is:

$$\frac{\text{change in } y}{\text{change in } x} = \frac{y_2 - y_1}{x_2 - x_1} = m$$

where m is a fixed number called a constant.

For the example of speed given above:

$$\text{speed} = \frac{\text{distance travelled}}{\text{time}}$$

The speed is the rate of change of distance travelled with respect to time.

In the same way a gradient, written:

$$\text{gradient} = \frac{\text{change in } y}{\text{change in } x} = m$$

can be thought of as the rate of change of y with respect to x.

The constant m is the rate of change of y with respect to x.

Gradient of a curve

A curve does not have a constant gradient – its direction is continuously changing, so its gradient will continuously change too. The gradient of a continuous curve with equation $y = f(x)$ at any point on the curve is defined as the gradient of the tangent to the curve at this point.

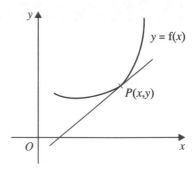

The **tangent** to a continuous curve at any point is a straight line which just touches the curve at this point. The gradient of the tangent to the curve $y = f(x)$ at the point $P(x, y)$ is the rate at which y is changing with respect to x.

8.2 Using geometry to approximate to a gradient

Look at this curve:

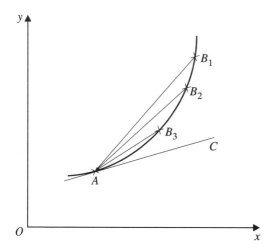

AC is the tangent to the curve at the point A.

A is a fixed point on the curve. Look at the chords AB_1, AB_2, AB_3, ... (A chord is a line joining two points on a curve.) For points B_1, B_2, B_3 ... that are closer and closer to A the sequence of chords AB_1, AB_2, AB_3 ... move closer to becoming the tangent AC.

The gradients of the chords AB_1, AB_2, AB_3 ... move closer to becoming the gradient of the tangent AC. So these gradients move closer to becoming the gradient of the curve at point A.

A numerical approach to rates of change

Here is how this idea can be applied to a real example. Look at the section of the graph of $y = x^2$ for $3 \leqslant x \leqslant 4$.

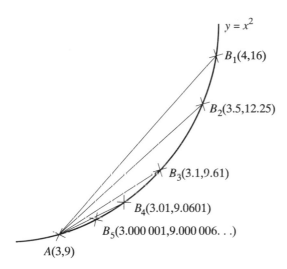

The gradient of the chord AB_1 is:

$$\frac{16 - 9}{4 - 3} = 7$$

Here are the gradients of the chords AB_1, AB_2, AB_3, AB_4 and AB_5 :

Chord	x changes from	y changes from	$\dfrac{\text{change in } y}{\text{change in } x} =$ gradient
AB_1	3 to 4	9 to 16	$\dfrac{16 - 9}{4 - 3} = 7$
AB_2	3 to 3.5	9 to 12.25	$\dfrac{12.25 - 9}{3.5 - 3} = 6.5$
AB_3	3 to 3.1	9 to 9.61	$\dfrac{9.61 - 9}{3.1 - 3} = 6.1$
AB_4	3 to 3.01	9 to 9.0601	$\dfrac{0.0601}{0.01} = 6.01$
AB_5	3 to 3.000 001	9 to 9.000 006 . . .	$\dfrac{0.000 006 . . .}{0.000 001} = 6$

(Notice that in this last case, the ratio is approximate.)

At the point (3.000 001, 9.000 006 . . .) the line AB_5 is virtually the same as the tangent to the curve at A. Its gradient is 6.

To convince yourself further using a numerical approach you could take a further series of points on the curve at $x = 2$, 2.5, 2.9, 2.99 and 2.999 999 (so that you are approaching $x = 3$ from below) and show that they also give gradients which progressively approach the value of 6.

This numerical approach shows that the gradient of the curve $y = x^2$ at the point $A(3,9)$ is 6.

8.3 A general approach to rates of change

Using a step-by-step approach similar to that in section 8.2 it is possible to find a numerical value for the gradient or rate of change at *any* point on a curve. The method can be generalised by taking a point $P(x, y)$ on the curve $y = f(x)$.

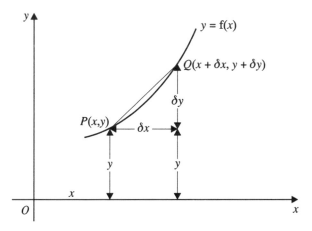

The point $Q(x + \delta x, y + \delta y)$ is very close to P on the curve. The small change from P in the value of x is δx and the corresponding small change in the value of y is δy. It is important to understand that δx is read as 'delta x' and is a *single symbol*. It is often called the **increment** (or change) in x and δy is called the increment in y.

The gradient of the chord PQ is:

$$\frac{(y + \delta y) - y}{(x + \delta x) - x} = \frac{\delta y}{\delta x}$$

As the equation of the curve is $y = f(x)$, the coordinates of P can also be written as $[x, f(x)]$ and the coordinates of Q as $[(x + \delta x),\ f(x + \delta x)]$.

The value of δx can be made as small as you like. The smaller the value of δx, the smaller the value of δy and the closer point Q will be to point P. As shown in section 8.2 the ratio $\dfrac{\delta y}{\delta x}$ approaches a definite limit as δx gets smaller and approaches zero. This limit is the gradient of the tangent at P which is the gradient of the curve at P. It is called the **rate of change of y with respect to x** at the point P.

This is denoted by $\dfrac{dy}{dx}$.

$$\frac{dy}{dx} = \lim_{\delta x \to 0} \left(\frac{\delta y}{\delta x} \right)$$

$$= \lim_{\delta x \to 0} \left[\frac{f(x + \delta x) - f(x)}{(x + \delta x) - x} \right]$$

$$= \lim_{\delta x \to 0} \left[\frac{f(x + \delta x) - f(x)}{\delta x} \right]$$

The symbol $\dfrac{dy}{dx}$ is called the **derivative** or the **differential coefficient** of y with respect to x. Read aloud it sounds like 'dee y by dee x'. In words $\lim\limits_{\delta x \to 0} \left(\dfrac{\delta y}{\delta x} \right)$ is: 'the limit of $\dfrac{\delta y}{\delta x}$ as δx tends to zero'. 'Tends to' is another way of saying 'approaches'.

If $y = f(x)$, you can also use the notation:

$$\frac{dy}{dx} = f'(x)$$

In this case f' is often called the **derived function** of f.

The procedure used to find $\dfrac{dy}{dx}$ from y is called **differentiating y with respect to x**.

Example 1

Find $\dfrac{\mathrm{d}y}{\mathrm{d}x}$ for the function $y = x^2$.

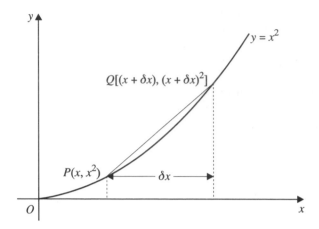

Taking P as (x, x^2) and a neighbouring point Q as $[(x + \delta x), (x + \delta x)^2]$ gives:

$$\frac{\mathrm{d}y}{\mathrm{d}x} = \lim_{\delta x \to 0}\left[\frac{(x + \delta x)^2 - x^2}{(x + \delta x) - x}\right]$$

$$= \lim_{\delta x \to 0}\left[\frac{x^2 + 2x\,\delta x + (\delta x)^2 - x^2}{\delta x}\right]$$

$$= \lim_{\delta x \to 0}\left[\frac{2x\,\delta x + (\delta x)^2}{\delta x}\right]$$

$$= \lim_{\delta x \to 0}(2x + \delta x)$$

As δx approaches zero ($\delta x \to 0$) the limiting value is $2x$. That is, the derivative of x^2 with respect to x is $2x$. (You could also say that the derived function of x^2 with respect to x is $2x$.) More often this is written:

$$\frac{\mathrm{d}y}{\mathrm{d}x} = 2x$$

For the curve $y = x^2$, we have generally $\dfrac{\mathrm{d}y}{\mathrm{d}x} = 2x$. So when $x = 3$, $\dfrac{\mathrm{d}y}{\mathrm{d}x} = 6$.

Look back at section 8.2 to see how the numerical approach used there agrees with the more general result used in example 1.

Example 2

Find $\dfrac{\mathrm{d}y}{\mathrm{d}x}$ for the function $y = \dfrac{1}{x}$.

For neighbouring points $P(x, y)$ and $Q\left(x + \delta x,\ \dfrac{1}{x + \delta x}\right)$:

$$\frac{\mathrm{d}y}{\mathrm{d}x} = \lim_{\delta x \to 0}\left[\frac{\left(\dfrac{1}{x + \delta x} - \dfrac{1}{x}\right)}{(x + \delta x) - x}\right] = \lim_{\delta x \to 0}\left[\frac{\left(\dfrac{x - x - \delta x}{x(x + \delta x)}\right)}{\delta x}\right]$$

So:
$$\frac{\mathrm{d}y}{\mathrm{d}x} = \lim_{\delta x \to 0}\left[\frac{-\delta x}{x\,\delta x(x + \delta x)}\right] = \lim_{\delta x \to 0}\left[\frac{-1}{x(x + \delta x)}\right]$$

The limit as δx approaches zero is:

$$\frac{\mathrm{d}y}{\mathrm{d}x} = \frac{-1}{x^2} = -x^{-2}$$

So for $y = \dfrac{1}{x}$ (or x^{-1}),

$$\frac{\mathrm{d}y}{\mathrm{d}x} = -\frac{1}{x^2} \quad (\text{or } -x^{-2})$$

8.4 A general formula for $\dfrac{\mathbf{d}y}{\mathbf{d}x}$ when $y = x^n$

If you extend the general results obtained in examples 1 and 2 for further functions of x such as x^3 and $x^{\frac{1}{2}}$, you will find that $\dfrac{\mathrm{d}y}{\mathrm{d}x} = 3x^2$ and $\dfrac{\mathrm{d}y}{\mathrm{d}x} = \frac{1}{2}x^{-\frac{1}{2}}$.

■ **In general, when $y = x^n$, where n is any real number:**

$$\frac{\mathrm{d}y}{\mathrm{d}x} = nx^{n-1}$$

Here are some other useful results which you can use. Their proofs are not shown.

■ **When $y = kx^n$ where k is a constant:**

$$\frac{\mathrm{d}y}{\mathrm{d}x} = nk\,x^{n-1}$$

■ **When $y = u + v$ where u and v are functions of x**

$$\frac{\mathrm{d}y}{\mathrm{d}x} = \frac{\mathrm{d}u}{\mathrm{d}x} \pm \frac{\mathrm{d}v}{\mathrm{d}x}$$

Example 3

Find $\dfrac{dy}{dx}$ for each of the following:

(a) $y = x^3 - x^7$ (b) $y = 2x^4 - 3$ (c) $y = \dfrac{1}{x^3} - \dfrac{2}{x}$

Apply the formula:

(a) $\dfrac{dy}{dx} = 3x^{3-1} - 7x^{7-1} = 3x^2 - 7x^6$

(b) $\dfrac{dy}{dx} = 2(4x^{4-1}) - 0 = 8x^3$.

(Notice that $3 = 3 \times 1 = 3x^0$. So $\dfrac{dy}{dx} = 3(0x^{0-1}) = 0$. In other words a constant such as 3 has no rate of change and its derivative is zero.)

(c) Write this first as $y = x^{-3} - 2x^{-1}$.

Then: $\qquad\qquad \dfrac{dy}{dx} = -3x^{-3-1} - 2(-1x^{-1-1})$

$$= -3x^{-4} + 2x^{-2}$$

$$= -\dfrac{3}{x^4} + \dfrac{2}{x^2}$$

Example 4

You may need to find the value of $\dfrac{dy}{dx}$ at a particular point.

Find the value of $\dfrac{dy}{dx}$ at the point where $x = 3$ on the curve whose equation is $y = (2x - 1)(3x + 2)$.

Multiply out the brackets giving:

$$y = 2x(3x + 2) - 1(3x + 2)$$
$$= 6x^2 + 4x - 3x - 2$$
$$= 6x^2 + x - 2$$

Differentiate with respect to x:

$$\dfrac{dy}{dx} = 6(2x^{2-1}) + 1x^{1-1} - 0$$
$$= 12x + 1$$

At the point where $x = 3$, $\dfrac{dy}{dx} = 36 + 1 = 37$

Exercise 8A

1 Find the y-coordinate of the point Q on the curve $y = x^3$ when the x-coordinate is

(a) 2 (b) 1.5 (c) 1.1 (d) 1.01 (e) 1.0001

The point P has coordinates $(1,1)$.

Find the gradients of the chords joining P to Q in each case. Use them to find an estimate for the gradient of the tangent to the curve at P.

2 Repeat question 1 for the curve with equation

(a) $y = x^{-2}$ (b) $y = \sqrt{x}$ (c) $y = x^{-\frac{1}{3}}$.

3 In each of the following y is given as a function of x. Find the derived function $\dfrac{dy}{dx}$.

(a) x^4 (b) x^{-3} (c) -2

(d) $3x$ (e) $4x^3$ (f) $2x^{-5}$

(g) $\dfrac{1}{2x}$ (h) $\dfrac{3}{2x^2}$ (i) $x^2 - x^{2}$

(j) $3x^{-1} - 2x^{-2}$ (k) $(x-1)(x+2)$ (l) $x(x^2 - 3)$

(m) $x^2(x^{-1} - x^{-2})$ (n) $(2x^2 - 3)^2$ (o) $\dfrac{x^3 - 1}{2x}$

4 Find the gradient of the curve with equation $y = f(x)$ at the point where $x = a$ when:

(a) $f(x) = x^6$, $a = -1$ (b) $f(x) = -3x^2$, $a = 2$

(c) $f(x) = 1 - 3x^2$, $a = \frac{1}{2}$ (d) $f(x) = \dfrac{1}{\sqrt{x}}$, $a = 4$

(e) $f(x) = \sqrt{(12x)}$, $a = \sqrt{3}$ (f) $f(x) = x^2(2x - 1)$, $a = -1$

(g) $f(x) = \dfrac{x - 4}{x^2}$, $a = -2$

5 Here is a sketch of the curve $y = x^2 - 2x$, which cuts the x-axis at the origin O and the point A, and passes through the point $B(1, -1)$. Find the gradient of the curve at the points A, B and O.

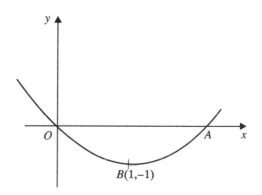

6   The points P, Q and R on the curve $y = 1 - x^3$ have x-coordinates 2, -1 and -2. The curve crosses the x-axis at S. Find the gradient of the curve at P, Q, R and S.

7   Find the y-coordinate and the value of $\dfrac{dy}{dx}$ at the point P whose x-coordinate is -1 on the curve $y = 2x^2 + \dfrac{1}{x}$.

8   The curve $y = 100x - x^2$ represents the path of an arrow fired from the origin O and landing on the horizontal ground again at the point A. Find the coordinates of A and the gradient of the path of the arrow at O and A. Find also the coordinates of the point H at which $\dfrac{dy}{dx} = 0$. Explain what is happening to the arrow at this point.

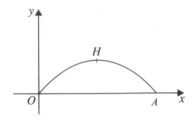

8.5 Exponential functions

Consider the functions given by the equations $y = 2^x$ and $y = 3^x$ in which the independent variable x is the index. Let's start by sketching their graphs for values of x from -2 to 2.

x	-2	-1	0	1	2
2^x	0.25	0.5	1	2	4
3^x	0.11	0.33	1	3	9

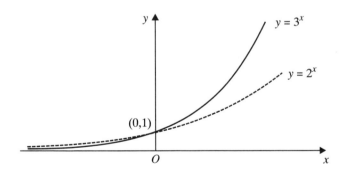

These curves are called **exponential** curves. They have a distinctive shape. The curve $y = a^x$ (where a is a constant greater than 1) increases steadily, passing through (0,1) as shown. The curve $y = a^{-x}$ decreases steadily, passing through (0,1), and is the reflection of the curve $y = a^x$ in the y-axis.

There is more about exponential curves on p. 85–6.

The gradient at any point $P(x,y)$ on $y = a^x$.

From the definition of $\dfrac{\mathrm{d}y}{\mathrm{d}x}$, the derivative of $y = a^x$ is:

$$\frac{\mathrm{d}y}{\mathrm{d}x} = \lim_{\delta x \to 0}\left(\frac{a^{x+\delta x} - a^x}{x + \delta x - x}\right) = \lim_{\delta x \to 0}\left(\frac{a^x a^{\delta x} - a^x}{\delta x}\right)$$

Here the rule of indices $t^{p+q} = t^p t^q$ has been used. (There is more about this rule on p. 81.)

Factorising the numerator and taking out the factor a^x gives:

$$\frac{\mathrm{d}y}{\mathrm{d}x} = a^x \lim_{\delta x \to 0}\left(\frac{a^{\delta x} - 1}{\delta x}\right)$$

If this derivative does approach a limit as δx approaches 0, the limit will depend in each case on the particular value of a.

For the curve $y = 2^x$ (where $a = 2$), home in on the limit by taking successively smaller values for δx, say 0.1, 0.01, 0.001 and 0.0001. These give values for $\left(\dfrac{2^{\delta x} - 1}{\delta x}\right)$ of 0.718, 0.696, 0.693 and 0.693 to 3 decimal places.

Similarly, for the curve $y = 3^x$ (where $a = 3$) home in on the limit by taking successively smaller values of δx at 0.1, 0.01, 0.001 and 0.0001. These give values for $\dfrac{3^{\delta x} - 1}{\delta x}$ of 1.161, 1.105, 1.099 and 1.099 to 3 decimal places. For the two curves the approximate values for their derivatives are:

$y = 2^x$: $\qquad\qquad\qquad \dfrac{\mathrm{d}y}{\mathrm{d}x} = (0.693)\,2^x$

$y = 3^x$: $\qquad\qquad\qquad \dfrac{\mathrm{d}y}{\mathrm{d}x} = (1.099)\,3^x$

These results suggest that exponential functions have a derivative which is proportional to the function. That is, $\dfrac{dy}{dx} = ky$ for some arbitrary constant k. This is in fact true, but you do not need to know the general proof at this stage. You can find the value of k for any particular function from your calculator by using the button marked 'ln': ln is the natural logarithm. (There is more about natural logarithms on pages 87–8 and 177–9.) In particular, check that:

$$\ln 2 = 0.693\,147\ldots \quad \text{and } \ln 3 = 1.098\,612\ldots$$

Setting the constant $k = 1$ gives a special function whose derivative is equal to the function itself, that is $\dfrac{dy}{dx} = y$. This special function is called the **exponential function** and a has the value e. The number e is irrational – it cannot be written exactly as a decimal. (There is more about irrational numbers on p. 10.) The value of e to 6 decimal places is $2.718\,282\ldots$

The exponential function is written as e^x. Here is its graph together with the graphs of the functions 2^x and 3^x.

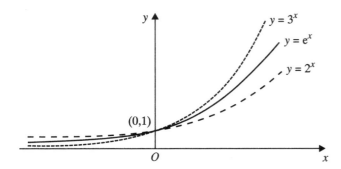

- **Remember that if $y = e^x$, then $\dfrac{dy}{dx} = e^x$.**

Most calculators have an e^x button so you can find the value of e^x at once for any value of x.

- **Memorise this result:**

When $y = e^{kx}$, $\dfrac{dy}{dx} = ke^{kx}$, where k is a constant

Example 5
Write down to 3 decimal places the values of:

(a) $e^{1.5}$ (b) e^{-1} (c) $e + e^{-2}$ (d) $\dfrac{1}{e} - \dfrac{1}{e^{1.3}}$

(a) $e^{1.5} = 4.482$

(b) $e^{-1} = 0.368$

(c) $e + e^{-2} = 2.718\,28\ldots + 0.135\,33\ldots = 2.854$

(d) $\dfrac{1}{e} - \dfrac{1}{e^{1.3}} = e^{-1} - e^{-1.3} = 0.3678\ldots - 0.2725\ldots$

$$= 0.095$$

Check through parts (a) to (d) of example 5 with your calculator to gain confidence in numerical work with the function e^x using the e^x and memory buttons.

Example 6
Find the gradient of the curve $y = e^{2x} - e^x$ at the points A, B and C where $x = -1$, 0 and 1. Give answers to 3 decimal places.

Differentiating with respect to x gives:

$$\frac{dy}{dx} = 2e^{2x} - e^x$$

This is the general expression for the gradient to the curve at the point with x-coordinate x.

At A, $x = -1$ and the gradient $= 2e^{-2} - e^{-1} = -0.331$

At B, $x = 0$ and the gradient $= 2e^0 - e^0 = 2 - 1 = 1$

At C, $x = 1$ and the gradient $= 2e^2 - e^1 = 12.060$

Exercise 8B

1 In each of the following, y is given as a function of x. Find $\dfrac{dy}{dx}$ in each case.

(a) e^{4x}

(b) e^{7x}

(c) $3e^{-2x}$

(d) $-e^{-5x}$

(e) $6e^{-\frac{x}{3}}$

(f) $e^{-2x} - 5$

(g) $4e^{-\frac{3x}{2}} + 2x$

(h) $e^{3x} - e^{-3x}$

(i) $e^x(e^x - 2)$

(j) $e^{-2x}(e^{2x} + e^{4x})$

(k) $x^2 - e^{3x}$

(l) $\dfrac{e^{-x} - e^{-3x}}{e^{2x}}$

(m) $e^{-x}(1 - e^{-2x})$

2   In parts (a) to (e) the equation of a curve is given, with the x-coordinate of a point P on the curve. Find the y-coordinate of P and the gradient of the curve at P. Give your answers to 3 significant figures.

(a) $y = e^{3x}$, $x = 2$ (b) $y = e^{\frac{x}{2}}$, $x = -2$

(c) $y = e^x + e^{-x}$, $x = 1$ (d) $y = x^2 + e^{\frac{2x}{3}}$, $x = 1$

(e) $y = e^{5x} - e^{-3x}$, $x = 0$

3   Draw on the same axes the graphs of $y = 2^x$ and $y = 3^{-x}$. Take values of x from -4 to 4 in each case. Write down the coordinates of the point of intersection T of the curves. Giving your answer to 2 decimal places estimate the gradient of each of the curves at T.

4   On separate diagrams make rough sketches of the curves whose equations are:

(a) $y = e^{-x}$ (b) $y = e^x - 2$ (c) $y = e^{x-2}$

On each diagram sketch the curve of $y = e^x$ with a dotted line so you can compare it with the curves (a) to (c).

5   Given that $y = ae^{-kx}$, where a and k are positive constants, prove that:

$$\frac{dy}{dx} + ky = 0$$

Sketch the graph of the function y when $a = 10$ and $k = 2$ for values of x in the interval $0 \leqslant x \leqslant 5$.

8.6 Functions of natural logarithms

Look at the curve $y = e^x$ shown in the diagram. You can see that the domain of the exponential function e^x is the set of real numbers (written $x \in \mathbb{R}$). The range of the exponential function is the set of positive real numbers (written $y \in \mathbb{R}^+$). There is more about set notation on p. 35.

The function e^x is a one-one function so an inverse function exists. You can draw a sketch of this inverse function by reflecting the curve $y = e^x$ in the line $y = x$ as shown. Notice that this inverse function has domain $x \in \mathbb{R}^+$ and that the function is not defined for zero or any negative number.

The inverse function of e^x is an important function in mathematics. It is called $\ln x$ (natural logarithm of x). The words logarithm and

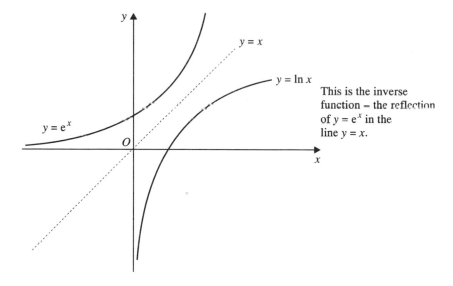

index mean the same thing. The properties of logarithms are described in detail in Book P2. At this stage you only need to be able to recognise the relation between the functions e^x and $\ln x$, the shapes of their graphs, and their derivatives.

The derivative of ln x

Consider the point $P(a, b)$ on the curve $y = \ln x$.

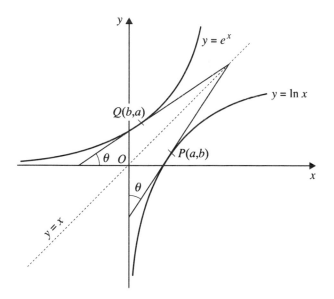

The tangent to the curve at P has gradient $\dfrac{dy}{dx}$. We want to find the derivative when $x = a$. In other words we want $\dfrac{dy}{dx}$ in terms of a.

Consider the point Q, which is the reflection of P in the line $y = x$. The point Q has coordinates (b, a) and lies on the curve $y = e^x$ because e^x is the inverse function of $\ln x$. For $y = e^x$, $\dfrac{dy}{dx} = e^x$ and the gradient of the tangent to the curve $y = e^x$ is e^b at $Q(b, a)$. But as Q is on the curve, we know that $a = e^b$. That is, the gradient of the curve $y = e^x$ at $Q(b, a)$ is a.

The curves $y = \ln x$ and $y = e^x$ are symmetrical about the line $y = x$. If the tangent to $y = e^x$ at (b, a) makes an angle θ with the x-axis, then the tangent to the curve $y = \ln x$ at (a, b) will make an angle θ with the y-axis. The gradient of the curve $y = e^x$ at Q is $\tan \theta = a$ (see p.67). Because of symmetry, the gradient of the curve $y = \ln x$ at the point $P(a, b)$ is $\tan(90° - \theta)$.

But: $$\tan(90° - \theta) = \frac{1}{\tan \theta} = \frac{1}{a}$$

So the gradient of the curve $y = \ln x$ at P is $\dfrac{1}{a}$. That is, for $y = \ln x$, $\dfrac{dy}{dx} = \dfrac{1}{a}$ at the point (a, b).

The point (a, b) can represent *any* point on the curve $y = \ln x$, so this is an important result:

- **When $y = \ln x$, $\dfrac{dy}{dx} = \dfrac{1}{x}$**

Example 7

Use the $\ln x$ button on your calculator to evaluate the following expressions giving your answers to 3 decimal places:

(a) $\ln 3 + \ln 7$ (b) $\ln 3 + \ln(\tfrac{1}{3})$ (c) $\ln 12 - \ln 4$

(a) $\ln 3 + \ln 7 = 1.0986.. + 1.9459\ldots = 3.045$

(b) $\ln 3 + \ln(\tfrac{1}{3}) = 1.0986\ldots + (-1.0986)\ldots = 0$

(c) $\ln 12 - \ln 4 = 2.4849\ldots - 1.386\,29\ldots = 1.098(= \ln 3)$

Example 8

Given that $y = x^2 - 3\ln x$, find the value of $\dfrac{dy}{dx}$ at $x = 2$.

Differentiating with respect to x gives:

$$\frac{dy}{dx} = \frac{d}{dx}(x^2) - 3\,\frac{d}{dx}(\ln x) = 2x - \frac{3}{x}$$

At $x = 2$,

$$\frac{dy}{dx} = 4 - \tfrac{3}{2} = 2\tfrac{1}{2}$$

Exercise 8C

1   Differentiate these expressions with respect to x:
   (a) $2\ln x$ (b) $\ln x + 6$ (c) $4\ln x + x^{\frac{1}{2}}$
   (d) $\tfrac{1}{2}\ln x + \tfrac{1}{4}x^2$ (e) $\tfrac{1}{3}\ln x + 2e^{\frac{1}{2}x}$ (f) $-5\ln x$
   (g) $\ln x + 3x^2$ (h) $e^{-2x} - 5\ln x$ (i) $\dfrac{2}{x} - 3\ln x$
   (j) $4\ln x - 3x^{-1}$

2   Sketch the graph of the curve $y = 1 + \ln x$ for $x \geqslant 0$. The points A and B have x-coordinates 2 and 4. Giving your answers to 3 decimal places find:
   (a) the y-coordinates of A and B
   (b) the gradient of the curve $y = 1 + \ln x$ at A and at B.

3   Find the value of $\dfrac{dy}{dx}$ for the curve $y = f(x)$ at the point P whose x-coordinate is given in each case.

	$f(x)$	x-coordinate of P
(a)	$3\ln x$	1.5
(b)	$x^2 + \tfrac{1}{3}\ln x$	2
(c)	$-\ln x$	0.2
(d)	$e^x - 6\ln x$	1
(e)	$x^2 - \ln x$	0.8

4   Sketch on separate axes the graphs of:
   (a) $y = 2 + \ln x$ (b) $y = \ln(x + 2)$ (c) $y = 2\ln x$
   In each sketch, show the graph of $y = \ln x$ with a dotted line.
   Find the gradient of each curve at the point on the curve where $x = 1$. $\left(\text{You may assume that } \dfrac{d}{dx}[\ln(x + 2)] = \dfrac{1}{x + 2}.\right)$

5   For the exponential function $f: x \mapsto e^x + 1$, $x \in \mathbb{R}$:
   (a) find the inverse function f^{-1} in a similar form
   (b) sketch the graphs of f, f^{-1} and the line $y = x$ on the same diagram
   (c) find the derived functions of f and f^{-1}.

8.7 Increasing and decreasing functions

A function f which increases as x increases in the interval from $x = a$ to $x = b$ is called an **increasing function** in the interval (a, b). Similarly a function f which decreases as x increases in the interval from $x = c$ to $x = e$ is called a **decreasing function** in the interval (c, e).

Example 9
The function $f(x) = x^3$ is increasing for all real values of x except at $x = 0$. This means that the tangent to the curve at any point (except the origin O) makes an acute angle with Ox. Notice also that $\dfrac{dy}{dx} = 3x^2$ and that $3x^2 > 0$ for all real values of x except $x = 0$.

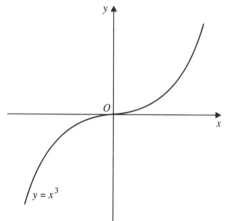

Example 10
Here is the graph of the function $f(x) = -(x - 1)^2$:

Like many other functions, this function is increasing in one interval (in the interval $x < 1$) and decreasing in another interval (the interval $x > 1$). At $x = 1$ the function is neither increasing nor decreasing. At this point it is said to be **stationary**.

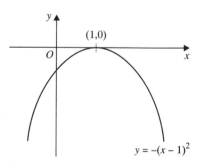

8.8 Turning points

On p.181 is the graph of a function $f(x)$ which is increasing in some intervals and decreasing in others. You can see that $f(x)$ decreases from A to B, then increases from B to C, and increases again from C to E.

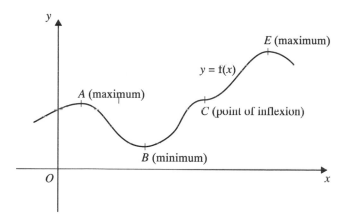

Suppose that the x-coordinates of A, B, C and E are a, b, c and e. Near point A, f(x) is increasing for $x < a$ and decreasing for $x > a$. At the point A, the curve reaches 'the top of a hill'. The value f(a) is called a **maximum** value of f. Also at A, f$'(a) = 0$; that is, the gradient of the curve is zero at A because the tangent to the curve at A is parallel to the x-axis.

Similarly at B, f$'(b) = 0$. The function f(x) is decreasing for $x < b$ and increasing for $x > b$. At the point B the curve reaches 'the lowest point'. The value f(b) is called a **minimum** value of f. Again, the tangent at B to the curve is parallel to the x-axis.

Between the points B and E, the curve is increasing, except at C where it levels out and has a tangent parallel to the x-axis. The point C is called an **inflexion** or a **point of inflexion**. At C, f$'(c) = 0$.

At any turning point on the curve $y = f(x)$, f$'(x) = 0$. The turning point may be a maximum, a minimum or a point of inflexion. You can find out what type a turning point is by considering the sign of f$'(x)$ near the turning point. Here are some more examples showing how to do this.

Example 11

Find the coordinates of the stationary points on the curve with equation $y = (x + 2)(x - 1)^2$. Sketch the curve, showing the stationary points and the coordinates of the points at which the curve meets the axes.

$$(x + 2)(x - 1)^2 = (x + 2)(x^2 - 2x + 1)$$
$$= x^3 - 2x^2 + x + 2x^2 - 4x + 2$$
$$y = x^3 - 3x + 2$$

Differentiating with respect to x gives:

$$\frac{dy}{dx} = 3x^2 - 3$$

Stationary points occur where $\frac{dy}{dx} = 0$, that is:

$$3(x^2 - 1) = 0$$
$$3(x - 1)(x + 1) = 0 \Rightarrow x = 1 \text{ or } x = -1$$

Substituting for x in $(x + 2)(x - 1)^2$ gives $y = 0$ or $y = 4$. The coordinates of the stationary points are $(1,0)$ and $(-1,4)$. The curve meets the x-axis when $(x + 2)(x - 1)^2 = 0$, that is, at $x = -2$ and at $x = 1$. The curve meets the y-axis when $x = 0$, that is at $(0,2)$.

Here is a sketch of $y = (x + 2)(x - 1)^2$:

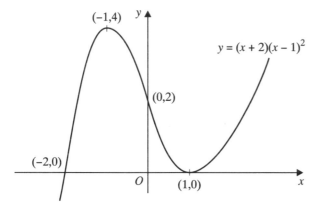

8.9 Using differentiation to solve practical problems

There are many practical problems that can be solved by mathematics where you want to find the maximum or minimum value of some quantity. This kind of problem can often be solved by differentiation.

When you first look at the problem, it may appear that the variable you want to maximise or minimise, y, depends on more than one other variable. You must use the data given in the problem to find a way of expressing y in terms of only one other variable x. Once $y = f(x)$ is set up, you can differentiate and test for stationary points as usual, to find the maximum (or minimum) value of y. Example 12 is typical of the kind of practical problems you will meet.

Example 12

A cylindrical tin, closed at both ends, is made of thin sheet metal. Find the dimensions of a tin like this that holds $1000\,\text{cm}^3$ and has a minimum total surface area.

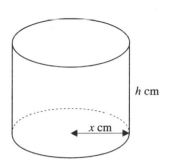

Let the tin have radius x centimetres and height h centimetres.

Then:
$$\text{volume} = \pi x^2 h = 1000 \Rightarrow h = \frac{1000}{\pi x^2}$$

The curved surface area is $2\pi xh$ square centimetres. The area of the two circular ends is $2\pi x^2$ square centimetres. Let the total surface area of the tin be y square centimetres.

Then:
$$y = 2\pi xh + 2\pi x^2$$

Both x and h vary. To solve the equation change it to an equation in two variables by substituting $h = \dfrac{1000}{\pi x^2}$ in the equation for y. This gives an equation in only two variables, x and y:

$$y = 2\pi x \left(\frac{1000}{\pi x^2} \right) + 2\pi x^2$$

So:
$$y = 2000x^{-1} + 2\pi x^2$$

The surface area y is now expressed as a function of x. Differentiating with respect to x gives:

$$\frac{\mathrm{d}y}{\mathrm{d}x} = -2000x^{-2} + 4\pi x$$

For a stationary value of y, $\dfrac{\mathrm{d}y}{\mathrm{d}x} = 0$.

So:
$$\frac{2000}{x^2} = 4\pi x$$

Rearranging gives:
$$x^3 = \frac{2000}{4\pi} = \frac{500}{\pi}$$

Taking the cube root, the value of x is 5.42. Substituting this value for x in the equation $h = \dfrac{1000}{\pi x^2}$ gives the value of h as 10.84. When $\dfrac{\mathrm{d}y}{\mathrm{d}x} = 0$, $x = 5.42$ and $h = 10.84$.

If you look at the sketch of the curve $y = \dfrac{2000}{x} + 2\pi x^2$, you can see that the only stationary point for $x > 0$ is a minimum. So the cylindrical tin of minimum surface area which has a volume of $1000\,\text{cm}^3$ has a radius of $5.42\,\text{cm}$ and a height of $10.84\,\text{cm}$.

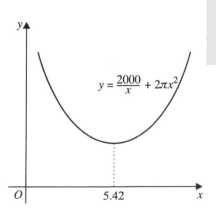

Example 13

Find the coordinates and the nature of the turning point on the curve $y = x^3 - 81 \ln x$.

$$\frac{dy}{dx} = 3x^2 - \frac{81}{x}$$

For a turning point, $\frac{dy}{dx} = 0$

That is:
$$3x^2 = \frac{81}{x}$$

So: $x^3 = 27$ and $x = 3$

At $x = 3$, $y = 27 - 81 \ln 3 \approx -62$

The turning point is at $(3, -62)$.

At $x = 2$, $\frac{dy}{dx} = 27 - 40.5 = -13.5 < 0$ and y is decreasing

At $x = 3$, we already know that y is stationary (i.e. the gradient of the tangent is zero)

At $x = 4$, $\frac{dy}{dx} = 48 - 20.25 = 27.75 > 0$ and y is increasing

The tangent gradients are negative for $x < 3$ and positive for $x > 3$. This is sufficient to conclude that the stationary value of y is a minimum because the sign of $\frac{dy}{dx}$ changes from $-$ to $+$ as x increases through the value 3. That is, the curve slopes downwards until the turning point is reached, has zero gradient at the turning point, and then slopes upwards.

We often show minimum values of y with the sketch

Maximum values are shown as

A point of inflexion for which $\frac{dy}{dx} = 0$ may be shown

as either or

Exercise 8D

1   Draw a sketch of the function $f(x) = -x^3$, $x \in \mathbb{R}^+$. Find $f'(x)$ and explain why f is a decreasing function.

2   For each of the following functions, find the set of values of x for which (i) the function is increasing, (ii) the function is decreasing. The domain of each is $x \in \mathbb{R}$. Sketch the function in each case.

   (a) $f(x) = (x - 2)^2$ (b) $f(x) = x^2 - 4$
   (c) $f(x) = -4x^2$ (d) $f(x) = x(2 - x)$

3   Find the least value of the function $f(x) = x^2 - 3x + 3$. Find the greatest value of the function $f(x) = 2x^2 + 6x - 6$.

4   Find the greatest value of $4x - x^2$ and state the value of x for which this occurs. If x takes all real values, find the range of the function $f(x) = 4x - x^2$.

5   Find the coordinates of the points on the following curves at which the gradient is zero. Describe the nature of each stationary point.

   (a) $y = x^2 + x^3$ (b) $y = x^3 - 3x + 2$
   (c) $y = x + x^{-1}$ (d) $y = x^2 + 27x^{-1}$
   (e) $y = (1 - x^2)(1 - 4x)$ (f) $y = x^3 - 3x^2 + 6$

6   Find the coordinates of the points on the curve $y = 2x^3 + 3x^2 - 12x + 6$ at which y has a stationary value. For each stationary value you find, say whether the value of y is a minimum or a maximum.

7   Show that the curve $y = 3x - x^3$ cuts the x-axis at three points. Find the turning points for this curve and so sketch the curve. Write all the information you have found on your sketch.

8   For the curves $y = 2x^3$ and $y = 4x - 5x^2$, find the values of x when the gradients of the two curves are equal.

9   A curve has equation $y = x^3 - 4x^2 + 4x + 3$.
   (a) Find the coordinates of the points on this curve at which the gradient is -1.
   (b) Find also the set of values of x for which y is decreasing.

10   For the curve $y = e^x - 4x$ find, giving your answers to 2 decimal places:
   (a) the coordinates of the stationary point

(b) the set of values of x for which y is increasing.

11   Given that $y = x - 2\ln x$, find the set of positive values of x for which y is increasing.

12   Given that $y = x^{-1}(x - 1)^2$ find the maximum and minimum values of y.

13   A stone is thrown vertically upwards and after t seconds its height y metres above its starting point is given by $y = 100t - 5t^2$. Calculate the greatest height reached by the stone and the value of t when this happens.

14   Given that $y = x^2 + 16x^{-1}$, find the maximum or minimum value of y and the value of x at which this occurs.

15   The sum of two variable positive numbers is 100. Let the numbers be x and $100 - x$ and let their product be y. Using differentiation find the maximum value of y.

16   The product of two variable positive numbers is 100. What is their least possible sum?

17   The organisers of a rave concert intend to use a large, flat expanse of land with a long, straight wall as one boundary. The rectangular enclosure is to be made from a total of 1000 metres of fencing on three sides with the wall for the fourth side. Given that the length of fencing on the side opposite the wall is $2x$ metres, show that the area y square metres of the enclosure is given by $y = 1000x - 2x^2$. Using differentiation find the maximum value of y.

18   A rectangular box with no lid is made from thin cardboard. The base is $2x$ centimetres long and x centimetres wide and the volume is 48 cubic centimetres. Show that the area, y square centimetres, of cardboard used is given by $y = 2x^2 + 144x^{-1}$. As x varies, find the value of x for which y is stationary and so find the minimum value of y.

19   A closed rectangular box is made of thin hardboard $3x$ centimetres long and x centimetres wide. The volume of the box is 288 cubic centimetres. Express the surface area, y square centimetres, of the box in terms of x and so find the value of x for which y is least.

20 A closed cylindrical olive oil tin is made of thin sheet metal of area 24π cm^2. Find the great possible capacity of the tin and the radius of the tin in this case.

21 A rectangular sheet of thin cardboard is 80 cm by 50 cm. A square of side x centimetres is cut away from each corner of the sheet which is then folded to form an open rectangular box of volume y cubic centimetres. Show that

$$y = 4000x - 260x^2 + 4x^3$$

Given that x varies, find the greatest volume of the box.

22 A rectangle of length x centimetres, where x varies, has a constant area of 12 square centimetres. Express the perimeter of the rectangle, y, in terms of x. Find the least possible value of y.

23 A circular ink stain gets larger as time goes on. Its radius x centimetres at time t seconds is given by $x = 0.1t^2$. Find the rate of change of x with respect to t when $t = 0.5$ and when $t = 1.5$. Express A square centimetres, the area of the stain at time t seconds, in terms of t and so find the rate of change of A with respect to t when $t - 0.5$ and when $t = 1.5$.

24 At time t seconds the radius x centimetres of an expanding sphere is given by $x = \frac{1}{2}t$. Express the volume, V cubic centimetres, and the surface, area, A square centimetres, in terms of t. From this find the rate of change of V and the rate of change of A with respect to t at the instant when $t = 3$.

25 The diameter of a closed cylindrical tin is equal to the height of the tin. Show that, for a fixed volume, a tin made to this specification requires the least amount of sheet metal.

SUMMARY OF KEY POINTS

1 The gradient of a curve $y = f(x)$ at the point P is defined to be equal to the gradient of the tangent at P to the curve.

2 The gradient of the tangent at P is the limit of the gradient of the chord PQ, where P is (x, y) and Q is $[x + \delta x, \ f(x + \delta x)]$.

3 $\quad \dfrac{dy}{dx} = \lim\limits_{\delta x \to 0} \left(\dfrac{\delta y}{\delta x} \right) = \lim\limits_{\delta x \to 0} \left(\dfrac{f(x + \delta x) - f(x)}{\delta x} \right) = f'(x)$

4 $\dfrac{dy}{dx}$ is called the **derivative** of y with respect to x, or the **differential coefficient of** y with respect to x. It is the **rate of change of y with respect x** at the point on the curve whose x-coordinate is x.

5 Memorise these standard formulae for derivatives:

$y = f(x)$	$\dfrac{dy}{dx} = f'(x)$ (k is a constant)
x^n	nx^{n-1}
kx^n	knx^{n-1}
e^x	e^x
e^{kx}	ke^{kx}
$\ln x$	x^{-1} or $\dfrac{1}{x}$
$u \pm v$	$\dfrac{du}{dx} \pm \dfrac{dv}{dx}$

6 If for the curve $y = f(x)$ in the interval $a < x < b$, $f'(x) > 0$, then f is an increasing function in the interval. Similarly, if $g'(x) < 0$, for $c < x < e$, then g is a decreasing function in the interval.

7 At a stationary point $f'(x) = 0$.

8 The nature of a turning point is determined like this:
If the sign of $f'(x)$ changes from $-$ to $+$ __/ the turning point is a minimum.
If the sign changes from $+$ to $-$ /‾‾\\ then the turning point is a maximum.
Points of inflexion occur when the sign of $f'(x)$ remains the same as x increases through the turning point.

Integration

9

Chapter 8 on differentiation is about the rate at which things change. Differentiation is the procedure used to find $\dfrac{dy}{dx}$ – the rate of change of a function y with respect to x.

You need a good understanding of the ideas in chapter 8 before working on chapter 9. This chapter looks at the reverse or inverse process to differentiation, called **integration**.

9.1 Integration as the inverse of differentiation

Chapter 8 on differentiation shows that if $y = x^3$, then $\dfrac{dy}{dx} - 3x^2$. That is, for a given function y the process of differentiation gives a unique derivative $\dfrac{dy}{dx}$. However, differentiating $y = x^3 - 5$ or $y = x^3 + 2.5$ gives the same result for $\dfrac{dy}{dx}$ in each case: $3x^2$. So if you know what $\dfrac{dy}{dx}$ is and you want to find y, there is more than one possible answer.

In this way differentiation and its inverse process, integration, are quite different. Differentiation produces a unique answer, but the result of integration is a whole set of solutions. If you know that $\dfrac{dy}{dx} = 3x^2$, then you can only say that

$$y = x^3 + C$$

where C is a constant. The solutions $y = x^3$, $y = x^3 - 5$ and $y = x^3 + 2.5$ are *all* members of this set of solutions. Here are graphs of these solutions:

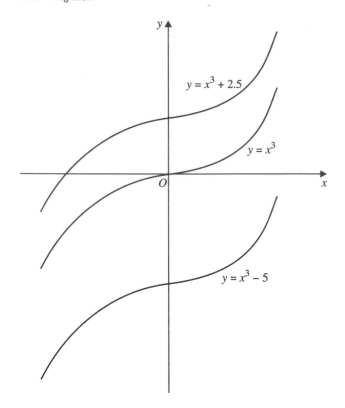

For different values of C, the equation $y = x^3 + C$ represents a set or family of curves. All of them are 'parallel' to one another, and no two curves ever pass through the same point.

This inverse process of moving from the differential relation $\dfrac{dy}{dx} = f'(x)$ to the general solution $y = f(x) + C$ is called **indefinite integration** – the constant C could take any value so the process is indefinite. A special symbol $\int$, looking like a stretched S, is used for integration. In this notation:

$$\int f'(x)\, dx = f(x) + C$$

In words this is: 'the integral of the function f' with respect to x equals the function f plus a constant'. As in the notation for differentiation, the dx stands for 'with respect to x'.

Working backwards from the standard formulae for differentiation in chapter 8 (p. 188) gives the following indefinite integrals, where k and C are constants.

■ $\displaystyle\int x\, dx = \tfrac{1}{2}x^2 + C$

■ $\displaystyle\int k\, dx = kx + C$

- $\int x^n \, dx = \dfrac{1}{n+1} \, x^{n+1} + C$, where $n \neq -1$

- $\int x^{-1} \, dx = \ln x + C$

- $\int e^x \, dx = e^x + C$

- $\int e^{kx} \, dx = \dfrac{1}{k} \, e^{kx} + C$

- $\int (u \pm v) \, dx = \int u \, dx \pm \int v \, dx$

Check all these results, except the last one, by differentiating the right-hand sides of the equations. **You need to memorise these results**.

Example 1
Find the following indefinite integrals:

(a) $\int x^7 \, dx$ (b) $\int x^{-3} \, dx$ (c) $\int x^{\frac{3}{2}} \, dx$

(a) Using the formula for $\int x^n \, dx$ with $n = 7$ gives:

$$\int x^7 \, dx = \tfrac{1}{8} x^8 + C$$

(b) Using the formula for $\int x^n \, dx$ with $n = -3$ gives:

$$\int x^{-3} \, dx = -\tfrac{1}{2} x^{-2} + C$$

(c) Using the formula for $\int x^n \, dx$ with $n = \frac{3}{2}$ gives:

$$\int x^{\frac{3}{2}} \, dx = \frac{1}{\frac{5}{2}} \, x^{\frac{5}{2}} + C$$

$$= \tfrac{2}{5} \, x^{\frac{5}{2}} + C$$

Example 2
Find $\int (6x^3 - 4x^{-1}) \, dx$

You need to find $6 \int x^3 \, dx - 4 \int x^{-1} \, dx$

$$= 6 \left(\frac{x^4}{4} \right) - 4 \ln x + C$$

$$= \frac{3x^4}{2} - 4 \ln x + C$$

Example 3

Find $\int (e^x + 1)^2 \, dx$

$$\int (e^x + 1)^2 \, dx = \int (e^{2x} + 2e^x + 1) \, dx$$

$$= \int e^{2x} \, dx + 2 \int e^x \, dx + \int 1 \, dx$$

$$= \tfrac{1}{2} e^{2x} + 2e^x + x + C$$

Exercise 9A

1 Find the following integrals:

(a) $\int x \, dx$ 　　(b) $\int x^4 \, dx$ 　　(c) $\int 6x^3 \, dx$

(d) $\int x^{-2} \, dx$ 　　(e) $\int 2x^{-4} \, dx$ 　　(f) $\int 9x^2 \, dx$

(g) $\int (4x + 5) \, dx$ 　　(h) $\int x(x - 1) \, dx$ 　　(i) $\int x^{-1}(x - x^2) \, dx$

(j) $\int (x + 1)^2 \, dx$ 　　(k) $\int (2 - x)^2 \, dx$ 　　(l) $\int \left(x - \dfrac{1}{x} \right)^2 \, dx$

(m) $\int (x^{-2} - x^2) \, dx$ 　　　　(n) $\int (1 + x^{-2})^2 \, dx$

(o) $\int (3x^5 - 2x^{-3}) \, dx$ 　　　　(p) $\int 15x^{-4} \, dx$

(q) $\int (3x - 2x^{-1})^2 \, dx$ 　　　　(r) $\int (3x - 2)(2x - 3) \, dx$

(s) $\int (5x^2 - 3)(5x^2 + 3) \, dx$ 　　(t) $\int x(\sqrt{x} - 2)^2 \, dx$

2 Integrate the following functions with respect to x:

(a) $x^{\frac{1}{2}}$ 　　(b) $x^{\frac{3}{5}}$ 　　(c) $x^{-\frac{1}{2}}$ 　　(d) $x^{-\frac{3}{4}}$

(e) $5x^{\frac{3}{2}}$ 　　(f) $\sqrt{(x^3)}$ 　　(g) $(\sqrt{x})^{-3}$ 　　(h) $4x^{\frac{1}{3}}$

(i) $3x^{\frac{5}{4}}$ 　　(j) $-\tfrac{1}{2}x^{-\frac{5}{4}}$ 　　(k) $(3x)^{\frac{1}{2}}$ 　　(l) $(4x)^{-\frac{1}{2}}$

3 Integrate the following with respect to x:

(a) e^{7x} 　　(b) e^{-3x} 　　(c) $2e^{2x}$ 　　(d) $\tfrac{1}{2}e^{\frac{x}{2}}$

(e) $e^{2x} - 3$ 　　(f) $6e^{-3x} - 3e^{3x}$ 　　(g) $(e^{2x} - 1)^2$

(h) $\dfrac{e^{3x} - e^x}{e^{2x}}$ 　　(i) $6x^2 + 4e^{-2x}$ 　　(j) $(3e^{-3x} + e^{-x})^2$

4 Find the following integrals:

(a) $\int (x^{-1} + 2)\, dx$

(b) $\int (2x - x^{-1})\, dx$

(c) $\int (2x^{-1} - x^{-2})\, dx$

(d) $\int \dfrac{3}{x}\, dx$

(e) $\int \dfrac{1}{3x}\, dx$

(f) $\int \dfrac{5}{8x}\, dx$

(g) $\int (3x^{-1} - 1)^2\, dx$

(h) $\int x(4 + x^{-2})\, dx$

(i) $\int \dfrac{3x^2 - 4}{2x}\, dx$

(j) $\int \dfrac{(3x - 2)(2x + 3)}{x^2}\, dx$

9.2 Boundary conditions

The process of getting to $f(x) + C$ from $\int f'(x)\, dx$ is called indefinite integration because of the inclusion of the constant C which could take any value. If you know a condition such as 'the value of y is b at $x = a$' then you can find C. The resulting equation for $f(x)$ in terms of x is then unique. A condition of this type is called a **boundary condition**. A boundary condition gives you extra information so that the value of the constant C can be found.

Example 4
A curve passes through the point $(2, -1)$. The gradient of the curve at any point is equal to $2 - 6x^2$. Find the equation of the curve.

You are told that $\dfrac{dy}{dx} = 2 - 6x^2$

Integrating gives:

$$y = 2x - 2x^3 + C$$

(Check by differentiating that the integration is correct.)

You know that $y = -1$ at $x = 2$. Substituting these values in the equation gives:

$$-1 = 4 - 16 + C$$

and: $\qquad\qquad\qquad C = 11$

So the equation of the curve is $y = -2x^3 + 2x + 11$.

Example 5

For all values of $t \geqslant 1$:

$$\frac{\mathrm{d}x}{\mathrm{d}t} = 2t^{-1} + 4t$$

Given that $x = 2$ at $t = 1$, find x in terms of t.

Integrating with respect to t gives:

$$x = 2\ln t + 2t^2 + C$$

where C is a constant.

At $t = 1$, $x = 2$ so:

$$2 = 2\ln 1 + 2 + C$$

But $\ln 1 = 0$ so $C = 0$.

The equation relating x and t is:

$$x = 2\ln t + 2t^2$$

9.3 Definite integrals

With indefinite integration we have:

$$\int f'(x)\,\mathrm{d}x = f(x) + C$$

where C is an arbitrary constant.

The **definite integral** $\displaystyle\int_a^b$ is defined by:

$$\int_a^b f'(x)\,\mathrm{d}x = \Big[f(x)\Big]_a^b = f(b) - f(a)$$

provided that f' is the derived function of f throughout the interval (a, b).

Notice the use of the **square brackets** around $f(x)$. This notation for definite integrals is standard. The numbers a and b are called the **limits** of the definite integral. In words $\displaystyle\int_b^a [g(x)]\,\mathrm{d}x$ is: 'the integral of $g(x)$ with respect to x between a and b'.

Example 6

Evaluate $\displaystyle\int_4^9 \sqrt{x}\,\mathrm{d}x$.

First write $\sqrt{x}$ in index form: $x^{\frac{1}{2}}$.

Using the formula for integrating x^n gives:

$$\int_4^9 \sqrt{x}\, dx = \left[\frac{1}{\frac{3}{2}}x^{\frac{3}{2}}\right]_4^9 = \frac{2}{3}\left[9^{\frac{3}{2}} - 4^{\frac{3}{2}}\right]$$

$$= \frac{2}{3}(27 - 8) = \frac{38}{3} = 12\frac{2}{3}$$

Example 7

Evaluate $\int_{-2}^0 (6e^{2x} - 3x)\, dx$.

The required integral is:

$$6\int_{-2}^0 e^{2x}\, dx - 3\int_{-2}^0 x\, dx = 6\left[\frac{e^{2x}}{2}\right]_{-2}^0 - 3\left[\frac{x^2}{2}\right]_{-2}^0$$

$$= 3[e^0 - e^{-4}] - 1.5[0 - 4]$$

$$= 3 - 3e^{-4} + 6$$

$$= 9 - 3e^{-4}$$

Note   Unless you are told to give your answer to so many decimal places in a question, the answer may be left in the form $9 - 3e^{-4}$.

Exercise 9B

1   In each of the following $\frac{dy}{dx}$ is given in terms of x, and $y = 1$ when $x = 0$. Find y in terms of x.

(a) $\frac{dy}{dx} = 2x - 1$ (b) $\frac{dy}{dx} = x^3 + x$ (c) $\frac{dy}{dx} = e^x + 2x$

(d) $\frac{dy}{dx} = e^{-2x}$ (e) $\frac{dy}{dx} = (3x - 4)^2$ (f) $\frac{dy}{dx} = e^{\frac{x}{2}}$

2   Evaluate each of the following definite integrals:

(a) $\int_0^2 3x^2\, dx$ (b) $\int_2^4 x^{-2}\, dx$ (c) $\int_1^4 \sqrt{x}\, dx$

(d) $\int_{-1}^1 (2x - 1)^2\, dx$ (e) $\int_{-1}^0 e^{-x}\, dx$ (f) $\int_2^5 \frac{3}{x}\, dx$

3   Show that $\int_0^2 (4 - x^2)^2\, dx = \frac{256}{15}$.

Evaluate the definite integrals in questions 4–7.

4 $\int_1^2 (x - 1)(x - 2)\, dx$ **5** $\int_4^9 (2x^{\frac{1}{2}} + 3x^{-\frac{1}{2}})\, dx$

6 $\displaystyle\int_{-1}^{2} x^3(4 - 5x)\, dx$ 7 $\displaystyle\int_{0}^{2} (e^x - 2)^2\, dx$

8 Find the value of $\displaystyle\int_{-1}^{1} x^n\, dx$ for $n = 6, 7, 8$ and 9.

9 Given that $\dfrac{dy}{dx} = x^3 + x^{-3}$ and that $y = 4$ when $x = 1$, find
   (a) y in terms of x (b) the value of y when $x = 2$.

10 For the curve $y = f(x)$, $f'(x) = 2x - \dfrac{x^2}{2}$. The curve passes through the point $(0,1)$. Find the value of $f(3)$.

11 Evaluate: (a) $\displaystyle\int_{4}^{9} x^{\frac{1}{2}}(2x - 3)\, dx$ (b) $\displaystyle\int_{-3}^{-1} \dfrac{(x - 1)}{x^4}\, dx$

12 Given that $\dfrac{dy}{dx} = 6x - 3x^2$, and $x \geqslant 1$, and also that $y = 6$ at $x = 1$, find the value of x for which y is greatest and so find the greatest positive value of y.

9.4 Finding the area of a region bounded by lines and a curve

Consider first two simple examples.

Example 8
Here is a sketch of the shaded region bounded by the line $y = h$, the ordinates $x = a$ and $x = b$ and the x-axis:

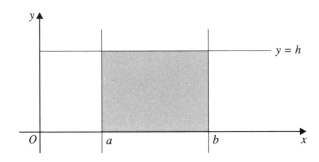

This region is a rectangle of length $(b - a)$ and width h. So the area of the region is $h(b - a)$. Notice also that:

$$\int_{a}^{b} h\, dx = \Big[hx\Big]_{a}^{b} = h\Big[x\Big]_{a}^{b}$$
$$= h(b - a)$$

The area of the region is $h(b - a)$.

Example 9

Here is a sketch of the shaded region bounded by the line $y = x$, the ordinates $x = a$ and $x = b$ and the x-axis.

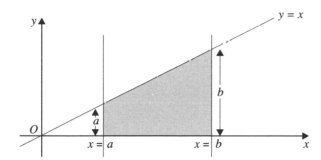

This region is a trapezium whose parallel sides are of length a and b. The perpendicular distance between these parallel lines is $(b - a)$ and the area of the trapezium is:

$$\frac{(a + b)}{2}(b - a) = \frac{b^2 - a^2}{2}$$

Notice also that:

$$\int_a^b x \, dx = \tfrac{1}{2}\left[x^2\right]_a^b$$
$$= \tfrac{1}{2}(b^2 - a^2)$$

The area of the region is $\tfrac{1}{2}(b^2 - a^2)$.

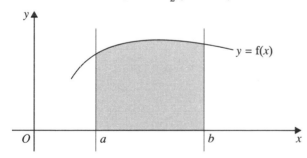

- **In general, the area of the region bounded by the curve $y = f(x)$, the ordinates $x = a$ and $x = b$ and the x-axis can be found by evaluating the definite integral $\int_a^b f(x) \, dx$, when it exists.**

Examples 8 and 9 show that this formula is valid for two simple cases when $f(x)$ is constant and when $f(x)$ is a linear function. A proof of the general case is not required in an advanced course. You are expected to assume the formula and to use it for the functions that you are able to integrate.

Example 10

Sketch a diagram to show the finite region bounded by the curve $y = x^{-1}$, the lines $x = 0.5$ and $x = 4$, and the x-axis. Find the area of this region.

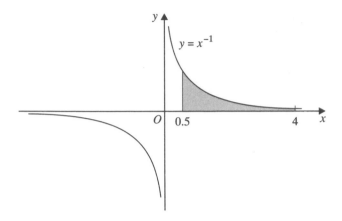

The sketch shows the region required.

The area of this region is:

$$\int_{0.5}^{4} x^{-1} \, dx = \Big[\ln x\Big]_{0.5}^{4}$$
$$= \ln 4 - \ln 0.5$$
$$= 2.08 \ (2 \text{ d.p.})$$

Example 11

Find the area of the finite region bounded by the curve $y = x^2 - 4$ and the x-axis.

The curve cuts the x-axis where $x^2 - 4 = 0$, that is at $(2,0)$ and $(-2,0)$. The curve cuts the y-axis at $(0,-4)$. Here is a sketch of the curve, which is similar to $y = x^2$ but with its lowest point at $(0,-4)$.

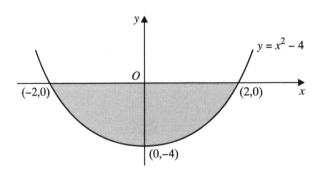

The shaded region is bounded by the x-axis and the curve.

The area of the shaded region is:

$$\int_{-2}^{2} (x^2 - 4) \, dx = \left[\frac{x^3}{3} - 4x \right]_{-2}^{2}$$
$$= \tfrac{8}{3} - 8 - (-\tfrac{8}{3} + 8)$$
$$= -\tfrac{32}{3}$$
$$= -10\tfrac{2}{3}$$

Notice that the negative sign means the region is below the x-axis. It does not mean that the area is negative. There is no such thing as a negative area!

Example 12

Here is a sketch of the curve $y = 2x - x^2$ which meets the line $y = -2x$ at the origin O and the point P. Determine the coordinates of P and so find the area of the shaded region.

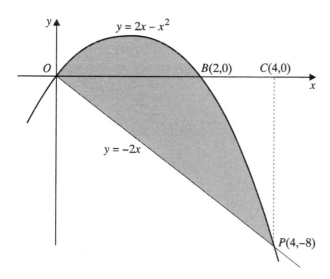

Find the coordinates of P by solving the simultaneous equations $y = -2x$ and $y = 2x - x^2$.

Replacing y by $-2x$ in the equation of the curve gives:

$$-2x = 2x - x^2$$

That is: $\qquad x^2 - 4x = 0$

Factorising gives: $\qquad x(x - 4) = 0$

So $x = 0$ at O. Also $x = 4$ and $y = -8$ at P.

The coordinates of P are $(4, -8)$.

Notice that the curve crosses the x-axis at the point $B(2,0)$. The perpendicular from P to the x-axis meets the x-axis at the point $C(4,0)$. You can now find the areas of three separate regions in the sketch. Here they are in separate sketches, labelled A_1, A_2 and A_3:

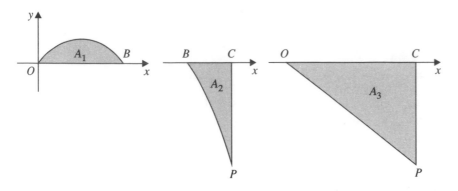

$$A_1 = \int_0^2 (2x - x^2)\, dx$$

$$= \left[\frac{2x^2}{2} - \frac{x^3}{3} \right]_0^2$$

$$= \tfrac{8}{2} - \tfrac{8}{3}$$

$$= 1\tfrac{1}{3}$$

$$A_2 = \int_2^4 (2x - x^2)\, dx$$

$$= \left[x^2 - \frac{x^3}{3} \right]_2^4$$

$$= (16 - \tfrac{64}{3}) - (4 - \tfrac{8}{3})$$

$$= 16 - \tfrac{64}{3} - 4 + \tfrac{8}{3}$$

$$= -6\tfrac{2}{3}$$

The area of triangle OCP is A_3.

$$A_3 = \tfrac{1}{2} \times 4 \times 8 = 16$$

The area of the shaded region is:

$$A_3 + A_1 - |A_2| = 16 + 1\tfrac{1}{3} - 6\tfrac{2}{3} = 10\tfrac{2}{3}$$

Notice that you have to find the areas of the regions above the x-axis and regions below the x-axis separately.

Exercise 9C

1   In each of the following cases find the area of the region
    bounded by the curve $y = f(x)$, the x-axis and the ordinates
    $x = a$ and $x = b$.

	$f(x)$	a	b
(a)	$6x^2 - 5$	1	3
(b)	$x^{\frac{5}{2}}$	1	4
(c)	$(1 - x^2)$	-1	1
(d)	e^{-x}	-3	0
(e)	$\dfrac{3}{4x}$	1	4
(f)	$3x^{-2}$	-3	-1
(g)	$e^{2x} + x^{\frac{1}{3}}$	0	3

2   A curve C passes through the origin O and its gradient at the
    point (x, y) is $(2x - 5)$. Find the coordinates of the other point
    B where the curve cuts the x-axis. Find the area of the region
    bounded by the line OB and the curve C.

3   Evaluate $\displaystyle\int_2^5 (2x + 1)\, dx$ and interpret this integral geometrically
    by considering the area of a trapezium.

4   Find the area of the region bounded by the x-axis, the curve
    $y = x - 8x^{-2}$ and the ordinates $x = 2$ and $x = 4$.

5   Find the area of the finite region bounded by the curve
    $y = 2 + x - x^2$ and the x-axis.

6   Find the area of the region bounded by the x-axis, the curve
    $y = (x - 6)^2$ and the ordinates $x = 2$ and $x = 5$. Explain the
    sign of the area obtained. Without further integration, state the
    value of $\displaystyle\int_2^5 (12x - x^2 - 36)\, dx$.

7   Sketch the curve $y = x^2$ and the line $y = 5x - 4$. Show that
    the curve and line intersect at the points P and Q whose
    x-coordinates are 1 and 4. Find the area of the finite region
    bounded by the line PQ and the curve.

8   Evaluate the integrals $\displaystyle\int_{-1}^0 x^3\, dx$, $\displaystyle\int_0^1 x^3\, dx$ and $\displaystyle\int_{-1}^1 x^3\, dx$.

    Interpret these results geometrically by sketching graphs.

9   Sketch the curve $y = (1 - x)(x - 4)$ and find the coordinates of the points where the curve intersects the x-axis. Shade in the regions whose areas are represented by $\int_1^4 (1 - x)(x - 4)\,dx$ and $\int_4^5 (1 - x)(x - 4)\,dx$. Without actually integrating, explain why:

$$\int_4^5 (1 - x)(x - 4)\,dx = \int_0^1 (1 - x)(x - 4)\,dx$$

10 Calculate the area of the finite region bounded by the curves $y = x^2$ and $y = \sqrt{x}$.
    (Hint: the curves are symmetrical about $y = x$ in the first quadrant.)

11 Find the area of the region bounded by the curve $y = x^2$ and the line $y = x$.

12 Sketch the curve $y = 2x^{-1}$ for $x \geqslant 3$. Evaluate the integral $\int_3^7 2x^{-1}\,dx$ and show in your diagram the region whose area is given by your evaluation of the integral.

13 The region R is bounded by the curve $y = x^2 + 2$, the x- and y-axes and the line joining the point $(2,6)$ to the point $(26,0)$, as shown in the diagram. Find the area of R.

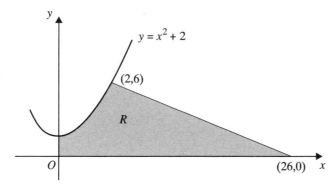

14 Find the area of the region bounded by the line $y = 2x$ and the curve $y = \sqrt{(20x)}$.

15 Show that the curve with equation $y = 2x^3 - 5x^2 - 4x + 3$ cuts the x-axis in three points A, B and C whose x-coordinates are $-1, \frac{1}{2}$ and 3.

    Find the area of the region bounded by the curve and the line segment AB.

16 Given that $\int_2^4 \left(3t^2 - 2t - \dfrac{k}{t^2}\right) dt = 40$, find the value of the constant k.

17 The line $y = 2$ meets the curve $y = 6 - x^2$ in two points A and B, as shown. Find the area of the shaded region.

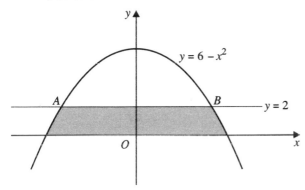

18 The region R is bounded by the curve $y = e^{-3x}$, the lines $x = -2$, $x = 3$ and $y = -4$, as shown. Calculate the area of R, giving your answer to 2 decimal places.

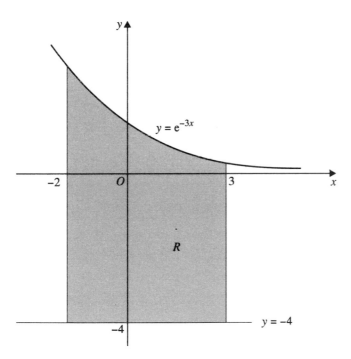

19 The points $A(-1, e^{-1})$ and $B(2, e^2)$ are on the curve $y = e^x$. Find to 2 decimal places, the area of the region bounded by the curve and the line segment AB.

20 For the curve $y = (x-1)(x+2)^2$ find:

(a) the coordinates of the turning points, distinguishing between a maximum and a minimum point

(b) the area of the region bounded by the curve, the y-axis and the positive x-axis.

SUMMARY OF KEY POINTS

1 The inverse operation to differentiation is integration.
2 Some standard indefinite integrals are:

$$\int f'(x)\,dx = f(x) + C$$

$$\int x^n\,dx = \frac{1}{n+1}x^{n+1} + C, \ n \neq -1$$

$$\int e^x\,dx = e^x + C$$

$$\int e^{kx}\,dx = \frac{1}{k}e^{kx} + C$$

$$\int \frac{1}{x}\,dx = \ln x + C$$

3 Definite integration:

$$\int_a^b f'(x)\,dx = \left[f(x)\right]_a^b = f(b) - f(a)$$

4 The area of the finite region between a curve $y = g(x)$, the x-axis and the ordinates $x = a$ and $x = b$, where $b > a$, is:

$$\int_a^b g(x)\,dx$$

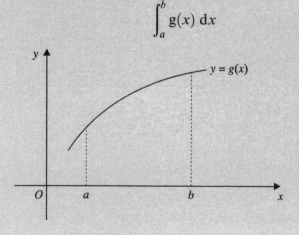

Numerical methods

10

No measurement is ever exact. If you measure the thickness of a piece of metal with a ruler, you might read the thickness as 3 mm. But if you measure the thickness of the same piece of metal with a micrometer (which measures more accurately), you might read the thickness as 3.4 mm. An even more accurate instrument may measure the thickness as 3.43 mm, and so on. None of these figures is completely accurate or precise. They are all **approximations** to the correct thickness. They are approximations because our measuring instruments, however good, can only measure to a certain degree of accuracy. So there will always be some degree of imprecision or error involved when you measure the thickness of the piece of metal.

Sometimes we deliberately introduce an error into measurements to make comparisons easier. For example, the population of one country may be quoted as 56 million and that of another country as 250 million. These are not accurate figures. The actual population of the first country as measured in the latest census could be 55 926 301 and the population of the other country could be 249 852 473. For some purposes figures given to the nearest million are just as helpful as the actual figures. For example, the rough figures of 56 million and 250 million tell you immediately that the population of one country is approximately five times the population of the other.

Remember that all measurements involve an error, and that many other figures may be counted precisely but quoted approximately so that they involve an error.

10.1 Absolute errors

Many numbers that cannot be quoted accurately are written to a given number of decimal places or significant figures. For example, 214.392 46 may be written as 214 (written to 3 significant figures) or 214.4 (written to 1 decimal place).

Example 1

If a number is given as 240 to the nearest ten then it lies between 235 and 245. It can be written as 240 ± 5.

Example 2

If a number is given as 3153 to 4 significant figures then it lies between 3152.5 and 3153.5. It can be written as 3153 ± 0.5.

When a number which is not written exactly is given as $(a \pm b)$, so that $(a + b)$ is the maximum possible value of the number and $(a - b)$ is the minimum possible value of the number, then b is called the **absolute error** in the number.

In example 1, the absolute error in 240 is 5. In example 2 the absolute error is 0.5.

Exercise 10A

The following numbers are given to the accuracy stated. For each number give its minimum possible value and its maximum possible value, and state the absolute error.

1 293   (3 significant figures)
2 160   (nearest ten)
3 1.124   (3 decimal places)
4 26 000   (nearest thousand)
5 120   (2 significant figures)
6 0.0126   (3 significant figures)
7 0.43   (nearest one hundredth)
8 8962   (nearest whole number)
9 1 324 168   (nearest whole number)
10 1 300 000   (nearest one hundred thousand)

10.2 Relative errors

Just giving an absolute error may make the error in a number seem more important than it actually is. For example, in a population count an error of 1000 people may sound a lot, but compared with the whole population of 56 000 000 it is actually rather small.

One way of giving an idea of the importance of the error is to relate the size of the error to the size of the number itself by finding the relative error. Sometimes the **relative error** is quoted instead of, or in addition to, the absolute error.

If e is the absolute error in the number x then the **relative error** is defined as $\frac{e}{x}$. So for the number 240 quoted to the nearest 10 in example 1 the relative error is:

$$\frac{5}{240} = \frac{1}{48}$$

For the number 3153 ± 0.5 it is

$$\frac{0.5}{3153} = \frac{1}{6306}$$

If the population of a village is mistakenly counted as 120 instead of 124 the absolute error is 4 and the relative error is $\frac{4}{124} \approx 0.032$. If the population of another village is mistakenly counted as 1250 instead of 1225 the absolute error this time is 25. This is five times the absolute error for the population of the first village, but the relative error is $\frac{25}{1225} \approx 0.020$ which is smaller than the relative error for the population of the first village. So the error in the count for the second village is less significant than the error for the first village.

Example 3

Calculate the relative error in 3150, which is given to 3 significant figures.

If 3150 is correct to 3 significant figures then its maximum possible value is 3155 and its minimum possible value is 3145. The number can be written 3150 ± 5, where 5 is the absolute error. The relative error is

$$\frac{5}{3150} \approx 0.0016$$

Exercise 10B

Calculate to 4 decimal places the relative errors in the following numbers given to the accuracy stated.

1 257   (3 significant figures)
2 10   (1 significant figure)

3 33.6   (1 decimal place)
4 0.19   (2 decimal places)
5 140   (nearest ten)
6 140   (nearest integer)
7 0.01   (1 significant figure)
8 353.2   (1 decimal place)
9 156   (nearest integer)
10 200   (nearest hundred)
11 200   (nearest ten)
12 26.63   (2 decimal places)

10.3 Calculating with approximations

When you do a calculation with two or more numbers the size of the error in the answer depends on the sizes of the errors in the numbers you start with. This section looks at the effect of adding, subtracting, multiplying and dividing approximate numbers.

Adding approximations

This is how to add the two numbers 3.7 and 5.6, each of which is given to two significant figures:

The maximum possible value of 3.7 is 3.75
The minimum possible value of 3.7 is 3.65
The maximum possible value of 5.6 is 5.65
The minimum possible value of 5.6 is 5.55

The maximum possible value of their sum is $3.75 + 5.65 = 9.4$ and the minimum possible value of their sum is $3.65 + 5.55 = 9.2$. So the true value of their sum lies between 9.2 and 9.4, and can be written as 9.3 ± 0.1.

The number 3.7 can be written as 3.7 ± 0.05 and the number 5.6 can be written as 5.6 ± 0.05.

Notice that: $\qquad\qquad 3.7 + 5.6 = 9.3$

and: $\qquad\qquad\qquad 0.05 + 0.05 = 0.1$

So the absolute error in the sum is 0.1 which is the sum of the errors in 3.7 and 5.6. It is $0.05 + 0.05$.

Example 4

Find the sum of 26 000 which is correct to 5%, 34 000 correct to 1% and 47 000 correct to 3%.

5% of 26 000 is 1300
1% of 34 000 is 340
3% of 47 000 is 1410

So the first number is $26\,000 \pm 1300$, the second number is $34\,000 \pm 340$, and the third is $47\,000 \pm 1410$.

The minimum sum of the three numbers is
$24\,700 + 33\,660 + 45\,590 = 103\,950$.

The maximum is $27\,300 + 34\,340 + 48\,410 = 110\,050$.

So the sum is $107\,000 \pm 3050$.

Now $107\,000 = 26\,000 + 34\,000 + 47\,000$. The absolute error in the sum, which is 3050, is the sum of the absolute errors $1300 + 340 + 1410$.

■ **When adding numbers, the value of the absolute error in the answer is the *sum* of the absolute errors in the numbers.**

General expression for the error in an addition

In general, if x_1 is the first number and e_1 its absolute error, and x_2 is the second number with absolute error e_2, then the maximum value of the first number is $x_1 + e_1$ and the minimum value of the first number is $x_1 - e_1$. The maximum value of the second number is $x_2 + e_2$ and the minimum value of the second number is $x_2 - e_2$.

The maximum value of the sum is:

$$(x_1 + e_1) + (x_2 + e_2)$$
$$= (x_1 + x_2) + (e_1 + e_2)$$

and the minimum value of the sum is:

$$(x_1 - e_1) + (x_2 - e_2)$$
$$= (x_1 + x_2) - (e_1 + e_2)$$

So the sum can be written:

$$(x_1 + x_2) \pm (e_1 + e_2)$$

This shows that the absolute error of the sum is $(e_1 + e_2)$ which is the sum of the two absolute errors.

Subtracting approximations

This is how to subtract 2.7 from 5.9 where each number is given to 2 significant figures:

The first number is 2.7 ± 0.05 and the second is 5.9 ± 0.05.

The largest value of the difference is obtained when the smallest number is taken from the largest number:

$$5.95 - 2.65 = 3.3$$

The smallest value of the difference is obtained when the largest value of 2.7 ± 0.05 is taken from the smallest value of 5.9 ± 0.05:

$$5.85 - 2.75 = 3.1$$

So the difference can be written 3.2 ± 0.1.

Notice that $3.2 = 5.9 - 2.7$ and that the absolute error is $0.1 = 0.05 + 0.05$.

■ **When subtracting, the value of the absolute error in the answer is the *sum* (not the difference) of the absolute errors in the two numbers.**

General expression for the error in a subtraction

This can be shown in general. If the first number is $(x_1 \pm e_1)$ and the second number is $(x_2 \pm e_2)$, then the maximum value of the difference is:

$$(x_1 + e_1) - (x_2 - e_2)$$
$$= x_1 + e_1 - x_2 + e_2$$
$$= (x_1 - x_2) + (e_1 + e_2)$$

The minimum value of the difference is:

$$(x_1 - e_1) - (x_2 + e_2)$$
$$= x_1 - e_1 - x_2 - e_2$$
$$= (x_1 - x_2) - (e_1 + e_2)$$

So the difference is $(x_1 - x_2) \pm (e_1 + e_2)$.

Multiplying approximations

If you multiply the numbers 2.3 and 5.7, each given to two significant figures, then the largest possible result is obtained from:

$$2.35 \times 5.75 = 13.5125$$

The smallest possible result is obtained from

$$2.25 \times 5.65 = 12.7125$$

This product can be written as: 13.1125 ± 0.4.

Now the absolute error in each of the original numbers is 0.05 so the relative error in 2.3 is:

$$\frac{0.05}{2.3} \approx 0.0217$$

and the relative error in 5.7 is:

$$\frac{0.05}{5.7} = 0.0088$$

The relative error in the product is

$$\frac{0.4}{13.1125} \approx 0.0305 = 0.0217 + 0.0088$$

■ **When multiplying, the relative error in the product is approximately the sum of the relative errors in the numbers multiplied.**

General expression for the error in a multiplication

To show this we again consider $x_1 \pm e_1$ and $x_2 \pm e_2$. The maximum value of the product is:

$$(x_1 + e_1)(x_2 + e_2) = x_1 x_2 + x_1 e_2 + x_2 e_1 + e_1 e_2$$

The minimum value of the product is:

$$(x_1 - e_1)(x_2 - e_2) = x_1 x_2 - x_1 e_2 - x_2 e_1 + e_1 e_2$$

Now if e_1 is small and e_2 is also small then $e_1 e_2$ is *very* small and can be neglected, so the maximum value of the product may be taken as $x_1 x_2 + x_1 e_2 + x_2 e_1$ and the minimum value of the product may be taken as $x_1 x_2 - x_1 e_2 - x_2 e_1$.

So the product is approximately $x_1 x_2 \pm (x_1 e_2 + x_2 e_1)$

The relative error in this product is:

$$\frac{x_1 e_2 + x_2 e_1}{x_1 x_2}$$

$$= \frac{x_1 e_2}{x_1 x_2} + \frac{x_2 e_1}{x_1 x_2}$$

$$= \frac{e_2}{x_2} + \frac{e_1}{x_1}$$

where $\dfrac{e_2}{x_2}$ is the relative error in the second number and $\dfrac{e_1}{x_1}$ is the relative error in the first number.

Dividing approximations

Here is how to divide the number 12 by 3.1, where each number is given to 2 significant figures.

Dividing 12 by 3.1 gives a maximum value of:

$$\frac{12.5}{3.05} \approx 4.0984$$

The minimum value is:

$$\frac{11.5}{3.15} \approx 3.6508$$

The quotient can thus be written:

$$3.8746 \pm 0.2238$$

So the relative error in the quotient is:

$$\frac{0.2238}{3.8746} \approx 0.0578$$

The relative error in 12 is:

$$\frac{0.5}{12} \approx 0.0417$$

and the relative error in 3.1 is

$$\frac{0.05}{3.1} \approx 0.0161$$

Notice that $0.0578 = 0.0417 + 0.0161$.

■ **For division the relative error in the quotient is approximately the *sum* of the relative errors in the original numbers.**

General expression for the error in a division

To show this generally, consider $(x_1 \pm e_1)$ and $(x_2 \pm e_2)$ again. The maximum value of the quotient is:

$$\frac{x_1 + e_1}{x_2 - e_2} = \frac{x_1\left(1 + \dfrac{e_1}{x_1}\right)}{x_2\left(1 - \dfrac{e_2}{x_2}\right)} = \frac{x_1}{x_2}\left(1 + \frac{e_1}{x_1}\right)\left(1 - \frac{e_2}{x_2}\right)^{-1}$$

The minimum value of the quotient is:

$$\frac{x_1 - e_1}{x_2 + e_2} = \frac{x_1\left(1 - \dfrac{e_1}{x_1}\right)}{x_2\left(1 + \dfrac{e_2}{x_2}\right)} = \frac{x_1}{x_2}\left(1 - \frac{e_1}{x_1}\right)\left(1 + \frac{e_2}{x_2}\right)^{-1}$$

Now you will have to take on trust that:

$$\left(1 - \frac{e_2}{x_2}\right)^{-1} \approx 1 + \frac{e_2}{x_2}$$

and

$$\left(1 + \frac{e_2}{x_2}\right)^{1} \approx 1 - \frac{e_2}{x_2}$$

(The derivation of these relationships is shown in Book P3.)

$$\text{The maximum value} \approx \frac{x_1}{x_2}\left(1 + \frac{e_1}{x_1}\right)\left(1 + \frac{e_2}{x_2}\right)$$

$$\approx \frac{x_1}{x_2}\left(1 + \frac{e_1}{x_1} + \frac{e_2}{x_2}\right)$$

if you ignore the very small expression: $\dfrac{e_1 e_2}{x_1 x_2}$

$$\text{The minimum value} \approx \frac{x_1}{x_2}\left(1 - \frac{e_1}{x_1}\right)\left(1 - \frac{e_2}{x_2}\right) \approx \frac{x_1}{x_2}\left(1 - \frac{e_1}{x_1} - \frac{e_2}{x_2}\right)$$

if you ignore $\dfrac{e_1 e_2}{x_1 x_2}$.

So the quotient can be written:

$$\frac{x_1}{x_2}\left[1 \pm \left(\frac{e_1}{x_2} + \frac{e_2}{x_1}\right)\right]$$

The relative error in the quotient is approximately:

$$\frac{\dfrac{x_1}{x_2}\left(\dfrac{e_1}{x_1} + \dfrac{e_2}{x_2}\right)}{\dfrac{x_1}{x_2}} = \frac{e_1}{x_1} + \frac{e_2}{x_2}$$

$\swarrow$ relative error in x_1 $\qquad \searrow$ relative error in x_2

Example 5

The number 13.5 is given to 3 significant figures and the numer 9.7 is given to 2 significant figures. For the sum of the two numbers find:
(a) the maximum value (b) the minimum value (c) the absolute error (d) the relative error.

13.5 to 3 significant figures can be written 13.5 ± 0.05

9.7 to 2 significant figures can be written 9.7 ± 0.05.

(a) The maximum value of the sum is $13.55 + 9.75 = 23.3$

(b) The minimum value of the sum is $13.45 + 9.65 = 23.1$

(c) The absolute error is $0.05 + 0.05 = 0.1$

(d) The relative error is $\dfrac{0.1}{13.5 + 9.7} = \dfrac{0.1}{23.2} \approx 0.0043$.

Example 6

The number 123 is given to 3 significant figures and the number 7.8 is given to 2 significant figures. For the quotient $\dfrac{123}{7.8}$ find: (a) the maximum value (b) the minimum value (c) the approximate relative error (d) the approximate absolute error.

The first number can be written 123 ± 0.5. The second number can be written 7.8 ± 0.05.

(a) The maximum value of the quotient is: $\dfrac{123.5}{7.75} \approx 15.94$

(b) The minimum value of the quotient is: $\dfrac{122.5}{7.85} \approx 15.61$

(c) The relative error in the quotient is approximately:

$$\frac{0.5}{123} + \frac{0.05}{7.8} \approx 0.0105$$

(d) The absolute error is given by:

$$\frac{e}{\frac{123}{7.8}} \approx 0.0105$$

$$e \approx 0.0105 \times \frac{123}{7.8}$$

$$e \approx 0.17$$

Exercise 10C

1 Given that the numbers are correct to the significant figures shown, find the maximum value and the minimum value of each of the following:

(a) $13.2 + 1.3$ (b) $13.2 - 1.3$ (c) 13.2×1.3

(d) $\dfrac{13.2}{1.3}$ (e) $0.47 - 0.13$ (f) 18×11

(g) $\dfrac{8.6}{4}$ (h) $15.1 + (6.7 \times 3.9)$

(i) $(5 \times 6.3) + (9 \times 4.4)$ (j) $12.5 - 6.2$

(k) $\dfrac{8.4 + 14.9}{1.7 + 0.5}$ (l) $(5.163)^2$

2   Given that the numbers are correct to the significant figures shown, find the absolute and relative errors in each of the following. Give the relative errors to 3 significant figures where appropriate.

(a) $16.2 + 0.035$ (b) $1216 + 1.27$ (c) $9.123 - 0.06$

(d) $183 - 47$ (e) $1080 + 2.46 - 13.8$

(f) $247 - 76.2 - 13.1$ (g) 19.2×3.4 (h) 6.7×0.3

(i) $\dfrac{19.2}{3}$ (j) $\dfrac{184}{12}$ (k) $(6.1 + 9.4) \times 13$

(l) $(8.2 + 3.1) \times (172 + 0.4)$ (m) $\dfrac{21.2 + 150}{19.8 - 7.3}$

3   The length of a rectangle is measured as 8.9 cm to 2 significant figures. Its width is 3.7 cm to 2 significant figures. Calculate:
(a) the maximum and the minimum possible values of the perimeter of the rectangle (b) the maximum and minimum possible values of the area of the rectangle.

4   Write down the value of $\sqrt{2}$ to 3 significant figures and the value of $\sqrt{5}$ to 2 significant figures. Using these approximations find the absolute error in:
(a) $\sqrt{2} + \sqrt{5}$ (b) $\sqrt{10}$
Find also the relative error when using these approximations of:
(c) $\sqrt{2} + \sqrt{5}$ (d) $\sqrt{10}$

5   The triangle ABC has a right-angle at B. The length of the side *AB* of the triangle is measured as 3.7 cm to 2 significant figures and the length of the side *BC* is measured as 5.7 cm to 2 significant figures. Calculate:
(a) the maximum possible area of the triangle
(b) the minimum possible area of the triangle
(c) the length *AC* to an appropriate degree of accuracy.
Use your answers to calculate
(d) the maximum possible perimeter
(e) the minimum possible perimeter.

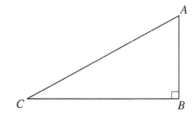

10.4 Finding the roots of f(x) = 0

Chapter 3 shows how to solve linear equations (such as $4x - 2 = 0$) and quadratic equations (such as $3x^2 - 2x - 1 = 0$). For polynomials of degree higher than two it is not always possible to find a factor, or factors. So trying to factorise the left-hand side of a polynomial equation is not a good way of finding the roots of the equation. This is particularly true when the roots are not integers, as trying to guess factors that are decimals is a very difficult exercise. There is also no simple algebraic method of finding the roots of equations which involve e^x, $\ln x$, $\sin x$, $\cos x$, etc. These are sometimes called **transcendental** equations.

For these types of equations you have to find out approximately where the root lies. Then you can use numerical methods to find closer and closer approximations to the root. Book P1 shows how to find the approximate locations of the roots of such equations. Other books in this series show how to use numerical methods to find closer approximations to the root.

Consider the function $f(x) \equiv x^2 - 3x - 10$. If you put $y = f(x)$, so that $y = x^2 - 3x - 10$, you can draw the graph of the function. It looks like this:

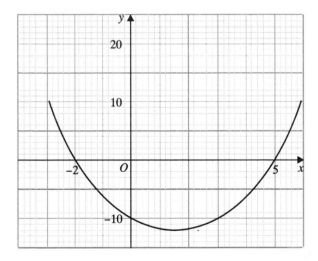

To find the roots of the equation $x^2 - 3x - 10 = 0$ you need to find points on the curve $y = x^2 - 3x - 10$ at which $y = 0$. Now on the graph the line $y = 0$ is the x-axis. So for the roots of the equation $x^2 - 3x - 10 = 0$ you need to find where the graph of $y = x^2 - 3x - 10$ cuts the x-axis. The graph cuts the x-axis at $x = -2$ and $x = 5$ so the roots of the equation are -2 and 5.

This procedure is one that works in general. To find the roots of $f(x) = 0$ first consider the graph of $y = f(x)$. It might look like this:

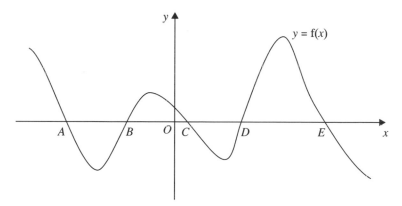

But for the equation $f(x) = 0$ you need to consider the curve $y = f(x)$ together with the line $y = 0$ (the x-axis). In this case the equation $f(x) = 0$ has five real roots which are given by the x-coordinates of the points A, B, C, D and E. It would be very time-consuming and impractical to draw a graph each time you wish to find the roots of an equation $f(x) = 0$ which you cannot solve algebraically. To find a better method, let's look at the graph of $y = f(x)$ to see what happens near a root.

To the left of point A on the graph the values of y are positive because the curve lies above the x-axis at these points. Just to the right of point A the values of y for points on the curve are negative because the curve is below the x-axis at these points. Just to the left of point B the values of y for points on the curve are negative and just to the right they are positive.

So the values of y for points on the graph of $y = f(x)$ *change sign* from positive to negative or negative to positive as the curve crosses the x-axis. This is when it passes through a root of the equation $f(x) = 0$. The graph shows that the same is true for points C, D and E.

In general, if you can find two values of x, one for which $f(x)$ is positive and one for which $f(x)$ is negative, then you know that the curve of $y = f(x)$ must have crossed the x-axis and so must have passed through a root of the equation $f(x) = 0$. So if $f(x_1) > 0$ and $f(x_2) < 0$ you know that there is at least one value of x lying between x_1 and x_2 which is a root of $f(x) = 0$. Similarly, if $f(x_3) < 0$ and $f(x_4) > 0$ there has been change of sign so at least one root of $f(x) = 0$ must lie between x_3 and x_4.

The only times this piece of detection does not work is where the graph looks like one of the following:

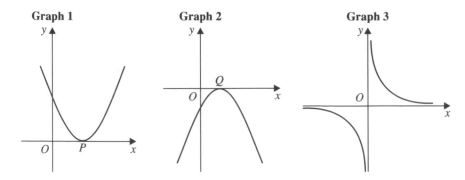

Graph 1　　　　　　Graph 2　　　　　　Graph 3

In Graph 1 there is a root of f(x) = 0 at P, but f(x) > 0 to the left of P and f(x) > 0 to the right of P.

Similarly, in Graph 2　f(x) < 0 to the left of Q and f(x) < 0 to the right of Q.

In Graph 3, the curve contains a **discontinuity** – a point at which the curve does not move continuously from one y value to the next. The curve never crosses the x-axis so no root exists. However, just to the left of the y-axis f(x) < 0 and just to the right of it f(x) > 0.

In the first two graphs the root cannot be detected by looking for a change in the sign of the value y. In Graph 3 you may think wrongly that a finite root exists because f(x) < 0 for x < 0 and f(x) > 0 for x > 0. However, the method does work in the overwhelming majority of cases.

Example 7
Show that $3 + 4x - x^4 = 0$ has a root somewhere between $x = 1$ and $x = 2$.

Let $f(x) \equiv 3 + 4x - x^4$

$f(1) = 3 + (4 \times 1) - 1^4 = 3 + 4 - 1 = 6 > 0$

$f(2) = 3 + (4 \times 2) - 2^4 = 3 + 8 - 16 = -5 < 0$

Because $f(1) > 0$ and $f(2) < 0$ there is a root of $3 + 4x - x^4$ between $x = 1$ and $x = 2$.

Example 8
Show that $\ln(1 + x) = e^{-x} + 1$ has a root near $x = 2$.

Rewrite the equation $\ln(1 + x) = e^{-x} + 1$ as:

$$\ln(1 + x) - e^{-x} - 1 = 0$$

and let:　　　　　　　$$f(x) \equiv \ln(1 + x) - e^{-x} - 1$$

$$f(1.5) = \ln 2.5 - e^{-1.5} - 1$$
$$\approx 0.916 - 0.223 - 1$$
$$= -0.307 < 0$$

$$f(2.5) = \ln 3.5 - e^{-2.5} - 1$$
$$\approx 1.253 - 0.082 - 1$$
$$= 0.171 > 0$$

As $f(1.5) < 0$ and $f(2.5) > 0$ there is a root of $\ln(1 + x) = e^{-x} + 1$ somewhere between $x = 1.5$ and $x = 2.5$.

Exercise 10D

1   Show that the equation $x^3 - 12x + 7 = 0$ has one negative real root and two positive real roots. Show that one of the roots lies between $x = \frac{1}{2}$ and $x = 1$.   [L]

2   Show that $x^3 - x + 3 = 0$ has a root between $x = -3$ and $x = +3$.

3   Given that $f(x) \equiv 3 + 4x - x^4$, show that the equation $f(x) = 0$ has a root $x = a$, where a is in the interval $1 \leqslant a \leqslant 2$.   [L]

4   Show that $x^3 + 2x - 4 = 0$ has a root in the vicinity of $x = 1.2$.

5   Show that a root of $x^3 + 8x - 28 = 0$ lies near 2.

6   Show that the equation $\sin x - \ln x = 0$ has a root lying between $x = 2$ and $x = 3$. Given that this root lies between $\dfrac{a}{10}$ and $\dfrac{a+1}{10}$, where a is an integer, find the value of a.   [L]

7   Show graphically or otherwise that the equation $x \ln x = 1 + x$ has only one real root and prove that this root lies between 3.5 and 3.8.   [L]

8   Show by means of a sketch that the equation $x^2 - 1 = e^{\frac{x}{2}}$ has three real roots. Show that one root lies in the interval $(-1.5, -1)$.   [L]

9   Show that the equation $e^x \cos 2x - 1 = 0$ has a root between 0.4 and 0.45.   [L]

10   Given that $f(x) \equiv 4x - e^x$, show that the equation $f(x) = 0$ has a root in the interval $0.3 \leqslant x \leqslant 0.4$.   [L]

11   It is given that $f(x) \equiv x - (\sin x + \cos x)^{\frac{1}{2}}$, $0 \leqslant x \leqslant \frac{3}{4}\pi$. Show that the equation $f(x) = 0$ has a root lying between 1.1 and 1.2.   [L]

12 By means of a sketch of the graphs $y = 8e^{-x}$ and $y = x^3$ show that the equation $8e^{-x} = x^3$ has exactly one real root. Denoting this real root by α find the integer n such that $n < \alpha < n + 1$.

[L]

13 $f(x) \equiv x - \pi(1 - \tan x)$, $|x| < \frac{1}{2}\pi$.

Calculate values for $f(0.6)$ and $f(0.7)$ and hence deduce that the equation $f(x) = 0$ has a root α, where $0.6 < \alpha < 0.7$. [L]

14 In the diagram, O is the centre of a circle, radius 10 cm, and the points A and B are situated on the circumference so that $\angle AOB = 2\theta$ radians. The area of the shaded segment is 44 cm^2. Show that $2\theta - \sin 2\theta - 0.88 = 0$.

Show further that a root of this equation lies between 0.9 and 1.

[L]

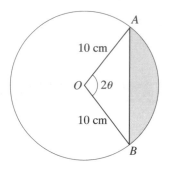

15 (a) Using the same axes, sketch the graphs of $y = e^x$ and $y = \dfrac{1}{x}$.

(b) Deduce the number of real solutions of the equation

$$e^x = \frac{1}{x}$$

and show that this equation may be written in the form

$$x - e^{-x} = 0$$

(c) Show that the equation $x - e^{-x} = 0$ has a root in the interval $0.5 < x < 0.6$. [L]

16 Using the same axes, sketch for $0 < x < 2\pi$ the graphs of $y = \sin x$ and $y = \ln x$.

x	1.5	2.0	2.5
$\sin x$			
$\ln x$			

Copy the table and use your calculator to complete the entries. By choosing further values of x and extending the table, find an estimate, to 1 decimal place, for the root of the equation

$$\sin x = \ln x$$

17 (a) Sketch, for $0 < x < \dfrac{\pi}{2}$, the curve $y = \tan x$. By using your sketch show that the equation $\tan x = \dfrac{1}{x}$ has one and only one root in $0 < x < \dfrac{\pi}{2}$.

(b) Show further that this root lies between 0.85 and 0.87.

[L]

SUMMARY OF KEY POINTS

1 When a number contains an error and the true value of that number lies in the range $x \pm e$, then e is called the **absolute error**.

2 For the number x with absolute error e, the quantity $\dfrac{e}{x}$ is the **relative error**.

3 If $X_1 = (x_1 \pm e_1)$ and $X_2 = (x_2 \pm e_2)$ then the absolute error in $X_1 + X_2$ is $e_1 + e_2$.

4 If $X_1 = (x_1 \pm e_1)$ and $X_2 = (x_2 \pm e_2)$ then the absolute error in $X_1 - X_2$ is $e_1 + e_2$.

5 If $X_1 = (x_1 \pm e_1)$ and $X_2 = (x_2 \pm e_2)$ then the relative error in $X_1 X_2 \approx \dfrac{e_1}{x_1} + \dfrac{e_2}{x_2}$.

6 If $X_1 = (x_1 \pm e_1)$ and $X_2 = (x_2 \pm e_2)$ then the relative error in $\dfrac{X_1}{X_2} \approx \dfrac{e_1}{x_1} + \dfrac{e_2}{x_2}$.

7 If $f(x_1) > 0$ and $f(x_2) < 0$ or if $f(x_1) < 0$ and $f(x_2) > 0$, then in general the equation $f(x) = 0$ has a root between x_1 and x_2.

Review exercise

1 The equation $x^4 - x^2 - 25 = 0$ has a root in the interval $(N, N+1)$, where N is an integer. Find the possible values of N.

2 Given that $y = 2x^2 - \dfrac{3}{x}$, find $\dfrac{dy}{dx}$ at $x = -1$. Show that the equation $y = 0$ has a root in the interval (1,2).

3 It is known that $-180° < x < 360°$ and $\tan x = \tan 35°$. Find the possible values of x.

4 Given that $\dfrac{dy}{dx} = \dfrac{12}{5x}$ and $x = 1$ when $y = 2$, find

   (a) y in terms of x (b) the value of y at $x = 10$.

5 Given that $y = e^{-2x} + 4$, find the value of $\dfrac{dy}{dx}$ and y when $e^{2x} = \frac{1}{2}$. Sketch the graph of $y = e^{-2x} + 4$.

6 The curve $y = p \sin x + q \cos x$ passes through the points (0,3) and $\left(\dfrac{\pi}{4}, 0\right)$. Find the values of p and q.

7 Evaluate $\displaystyle\int_{-1}^{2} (x-1)(2x+5)dx$.

8 Find the area of the finite region between the curve $y = 8x - x^2$ and the x-axis.

9 Find to 3 significant figures the gradient of the curve $y = e^{-x}$ at the point where $x = -1.5$.

10 Find the greatest value of $5x - 4x^2 - 1$.

11 Find in degrees to 1 decimal place in the interval $-180° \leqslant x \leqslant 180°$ the values of x for which
   (a) $\sin x = 0.7$ (b) $\cos x = 0.7$ (c) $\tan x = 0.7$

12 Find in radians to 2 decimal places in the interval $0 \leqslant x \leqslant 2\pi$ the values of x for which
   (a) $\sin x = -0.5$ (b) $\cos x = 0.8$ (c) $\tan x = 3$

13 In the diagram, O is the centre of the circle and AB is a chord. The radius of the circle is 10 cm and $\angle AOB - 150°$.
Calculate:

(a) the perimeter of the shaded region

(b) the area of the shaded region.

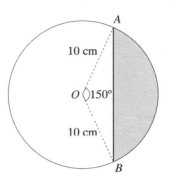

14

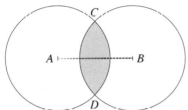

In the diagram, A and B are the centres of the circles, each with radius 10 cm. The circles intersect at C and D. Given that $\angle ACB = 2.33$ radians calculate:

(a) the area of the shaded region

(b) the length of AB to the nearest 0.1 cm.

15 Using intervals of 0.2 radians draw the graph of

$$y = 2\cos x + \sin x$$

in the interval $0 \leqslant x \leqslant 1.6$ and explain why the equation $y = 0$ has no root in this interval.

From your graph estimate:

(a) the greatest value of y

(b) the value of x for which $y = 1.5$.

16

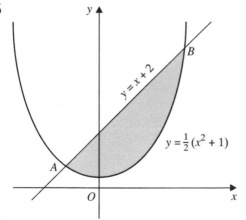

The diagram shows the curve $y = \frac{1}{2}(x^2 + 1)$ and the line $y = x + 2$, which meet at the points A and B.

(a) Find the coordinates of A and B.

(b) Determine the area of the shaded region.

(c) Find the range of the function f,

$$f : x \mapsto \tfrac{1}{2}(x^2 + 1), \ x \in \mathbb{R}$$

(d) Suggest a domain for which the function g,

$$g : x \mapsto \tfrac{1}{2}(x^2 + 1)$$

could have an inverse function.

17 Given that

$$\frac{\mathrm{d}x}{\mathrm{d}t} = 2t^2 - 10t + 12, \ t \geqslant 0$$

calculate:

(a) the values of t for which $\dfrac{\mathrm{d}x}{\mathrm{d}t} = 0$

(b) the value of x when $t = 4$, given further that $x = 2$ when $t = 1$.

18 Find the coordinates of the turning points on the curve whose equation is:

$$y = x^3 - 9x^2 + 24x$$

Sketch the curve for values of x from 0 to 5. Explain why the equation

$$x^3 - 9x^2 + 24x - 18 = 0$$

has 3 real roots.

Shade the region whose area is given by evaluating the integral

$$\int_2^4 (x^3 - 9x^2 + 24x)\mathrm{d}x$$

Evaluate the integral.

19 Find the area of the finite region bounded by the curve $y = 12 - x - x^2$ and the x-axis. Find also the range of the function f, defined by

$$f : x \mapsto 12 - x - x^2, \ x \in \mathbb{R}$$

20 Find the two smallest positive values of x where the curves $y = \sin x$ and $y = -\cos x$ intersect.

Give also the y-coordinates of these intersection points.

21 A rectangle has sides of 8.2 cm and 6.5 cm, where each length is given to 1 decimal place. Calculate:

(a) the largest possible perimeter of the rectangle

(b) the least possible area of the rectangle.

22 The length of a running track is 400 m to the nearest 10 m. An athlete takes 45 s, to the nearest second, to run around the track. Calculate the limits within which the average speed of the athlete must lie.

23 Written to 1 significant figure the number X is 5 and the number Y is 2. Prove that the actual value of $2X + 3Y$ lies between 13.5 and 18.5.

Find similar limits for XY and for $\dfrac{X}{Y}$.

24 The diagram shows a square card $ABCD$ of side 8 cm. From each of the corners is removed a square of side x cm, and the remainder is folded along each of the dotted lines to form an open tray of depth x cm and volume V cm^3.

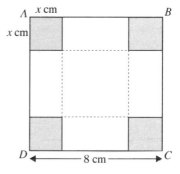

(a) Show that
$$V = 64x - 32x^2 + 4x^3$$

(b) Find $\dfrac{\mathrm{d}V}{\mathrm{d}x}$.

(c) Hence find the value of x for which $\dfrac{\mathrm{d}V}{\mathrm{d}x} = 0$.

(d) Show that this value of x gives a maximum value of V.

(e) Find the maximum value of V.

25 Sketch in the same diagram the graphs
$$y = \sin x$$
$$y = \sin 3x \quad \text{for } 0 \leqslant x \leqslant 90°$$

Hence, or otherwise, find the values of x for which

(a) $\sin 3x = 0.5$ (b) $\sin 3x = \sin x$.

26 For values of x from $-\dfrac{\pi}{2}$ to $\dfrac{\pi}{2}$, sketch the curves:
$$y = 0.5 - \sin x$$
$$y = 0.5 + 0.5 \cos 2x$$

Obtain the coordinates of the point P, at which these curves meet.

27 For a certain curve $\dfrac{dy}{dx} = 3x^2 - e^{2x}$ at any point P whose
   x-coordinate is x. Further, the curve passes through the point
   $(0,-1)$. Find the equation of the curve.

28 A solid circular cylinder is to have a volume of $16\pi\,\text{cm}^3$.
   Express the surface area $A\,\text{cm}^2$ of the cylinder in terms of the
   radius $x\,\text{cm}$ of its base. Hence find the value of x for which A
   is a minimum. Find also the minimum value of A.

29

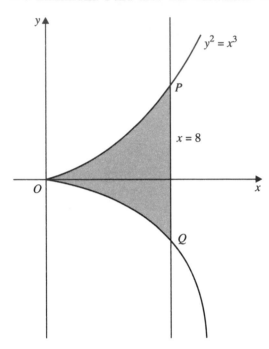

   The curve $y^2 = x^3$ is shown and the line $x = 8$ which meets
   the curve at P and Q. Find the length of the line PQ and the
   area of the shaded region.

30 The radius of a circular oil slick is $(4 + 5t^2)\,\text{cm}$ at time
   t seconds.
   Express the area $A\,\text{cm}^2$ at time t seconds in terms of t. Find
   the rate at which (a) the radius (b) the area of the slick is
   increasing at $t = 3$.

31 Find the area of the region bounded by the curve
   $y = 1 - x + x^2$, the lines $x = -1$ and $x = 2$ and the x-axis.

32 Find the area of the finite region bounded by the curve
   $y = 6x^2$ and the line $y = -5x - 1$.

33 Find the area of the finite region bounded by the curve
   $y = -2x^2 - 8x - 6$ and the x-axis.

34 Evaluate (a) $\displaystyle\int_1^{16} x^{-\frac{3}{4}}\,dx$ (b) $\displaystyle\int_2^5 \left(x - \frac{2}{x}\right)^2 dx$

(c) $\displaystyle\int_{-2}^1 e^{-\frac{1}{2}x}\,dx$

35 The curve $y = 2x - \dfrac{6}{x}$ cuts the positive x-axis at the point P.

At Q, R and S on the curve $x = 1,\ 3,\ 6$ respectively.

(a) Find the gradient of the line QR.

(b) Show that the gradient of the curve at P is equal to the gradient of QR.

(c) Find the area of the region bounded by the curve and the line RS.

(d) A root of the equation $y = 0$ lies in the interval $(N, N + 1)$. Find two values of N, where N is an integer.

36 The minute hand of a clock is of length $15\,cm$, measured to the nearest cm. Find the greatest possible distance moved by the tip of the minute hand between 1000 and 1530 on the same day.

37 In $\triangle ABC$ the angle B is $90°$ exactly. The lengths of AB and AC are $4\,cm$ and $5\,cm$ each to $0.1\,cm$. Calculate:

(a) the largest possible length of BC

(b) the limits between which the angle C must lie.

38 Explain the meaning of the terms (a) absolute error

(b) relative error.

(c) I measure a line and state that it is $5.7\,cm$ long to 2 s.f. Find (i) the absolute error (ii) the relative error.

39 The length, breadth and height of a cuboid are measured to the nearest mm as $18.7\,cm$, $11.6\,cm$ and $4.2\,cm$. Find the greatest possible volume of the cuboid.

40 A rectangular box without a lid has a square base of side $x\,cm$. The total internal surface area is $40\,cm^2$ and its walls and base are thin. Show that the volume $y\,cm^3$ is given by

$$y = \tfrac{1}{4}x(40 - x^2)$$

Given that x can vary, find the maximum volume of the box.

41 The triangle *ABC* is equilateral with each side of length 6 cm. With centre *A* and radius 6 cm, a circular arc is drawn joining *B* to *C*. Similar arcs are drawn with centres *B* and *C* and with radii 6 cm joining *C* and *A* to *B* respectively, as shown in the diagram. The shaded region *R* is bounded by the 3 arcs *AB*, *BC* and *CA*. Calculate, giving your answer in cm² to 3 significant figures:

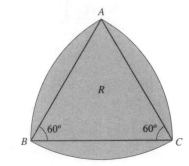

(a) the area of triangle *ABC*

(b) the area of *R*.   [L]

42

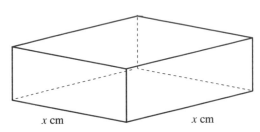

The diagram shows a rectangular cake-box, with no top, which is made from thin card. The volume of the box is 500 cm³. The base of the box is a square with sides of length *x* cm.

(a) Show that the area *A* cm² of card used to make such an open box is given by

$$A = x^2 + \frac{2000}{x}$$

(b) Given that *x* varies, find the value of *x* for which $\dfrac{\mathrm{d}A}{\mathrm{d}x} = 0$.

(c) Find the height of the box when *x* has this value.

(d) Show that when *x* has this value, the area of card used is least.   [L]

43 The diagram represents a circle of radius 10 cm and centre *C*. Points *A* and *B* are taken on the circumference of the circle so that $\angle ACB = 2$ radians. The shaded region *R* is bounded by the radii *CA* and *CB* and the major arc *ADB*, as shown. Calculate:

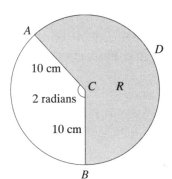

(a) the perimeter of *R*

(b) the area of *R*

(c) the area in cm² of $\triangle CAB$, giving your answer to 1 decimal place.   [L]

44 Evaluate $\displaystyle\int_{1}^{8} (x^{\frac{1}{3}} - x^{-\frac{1}{3}})\,dx$.

45 Given that $0 \leqslant x \leqslant \pi$, find the values of x for which

(a) $\sin 3x = 0.5$

(b) $\tan\left(x + \dfrac{\pi}{2}\right) = 1$. [L]

46 A curve C has equation $y = ax - x^2$, where a is a positive constant.

(a) Sketch C, showing clearly the coordinates of the points of intersection with the coordinate axis.

(b) Calculate the area of the finite region bounded by C and the x-axis, giving your answer in terms of a.

The lines with equations $x = \frac{1}{3}a$ and $x = \frac{2}{3}a$ intersect C at the points A and B, respectively.

(c) Find, in terms of a, the y-coordinates of A and B.

(d) Calculate the area of the finite region bounded by C and the straight line AB, giving your answer in terms of a.

[L]

47 The diagram shows a brick in the shape of a cuboid with base x cm by $2x$ cm and height h cm.

The total surface area of the brick is $300\,\text{cm}^2$.

(a) Show that $h = \dfrac{50}{x} - \dfrac{2x}{3}$.

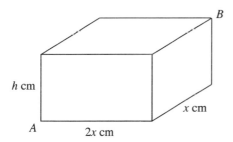

The volume of the brick is $V\,\text{cm}^3$.

(b) Express V in terms of x only.

Given that x can vary,

(c) find the maximum value of V. [L]

48

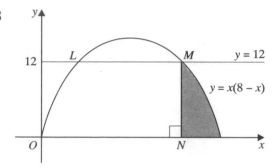

The diagram shows the curve C with equation $y = x(8 - x)$ and the line with equation $y = 12$ which meet at the points L and M.

(a) Determine the coordinates of the point M.

Given that N is the foot of the perpendicular from M onto the x-axis,

(b) calculate the area of the shaded region which is bounded by NM, the curve C and the x-axis. [L]

49 The curve with equation

$$y = 2 + k \sin x$$

passes through the point with coordinates $(\frac{\pi}{2}, -2)$. Find the value of k and the greatest value of y. [L]

50

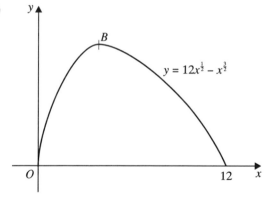

The diagram shows a sketch of the curve with equation

$$y = 12x^{\frac{1}{2}} - x^{\frac{3}{2}} \text{ for } 0 \leqslant x \leqslant 12.$$

(a) Show that $\dfrac{dy}{dx} = \dfrac{3}{2}x^{-\frac{1}{2}}(4 - x)$.

At the point B on the curve the tangent to the curve is parallel to the x-axis.

(b) Find the coordinates of the point *B*.

(c) Find the area of the finite region bounded by the curve and the *x*-axis.   [L]

Examination style paper

P1

1. Given that $p = \sqrt{2}$ and $q = \sqrt{3}$ express

   $$\frac{\sqrt{18} + \sqrt{12}}{\sqrt{8} - \sqrt{96}}$$ in terms of p and q in its simplest form.

   (6 marks)

2. A root of the equation $e^x - 7x = 0$ lies in the interval $N < x < N + 1$, where N is an integer. Find two values for N.

   (6 marks)

3. Find equations for the lines L_1 and L_2 where (*a*) L_1 passes through (3,2) and has gradient 3, (*b*) L_2 has gradient -2 and y intercept 13. Find the coordinates of the point of intersection of L_1 and L_2.

   (8 marks)

4. The function f is defined by

   $$f : x \mapsto 2x^3 - 3x^2 - 12x + 5, \ x \in \mathbb{R}.$$

   Find the set of values of x for which f is increasing.

   (8 marks)

5. (*a*) Find the solutions of the equation

   $$x^2 - 5x = 5,$$

   giving your answers to 2 decimal places.
   (*b*) Find the solution set of the inequality

   $$x^2 - 5x > 6.$$

   (9 marks)

6.

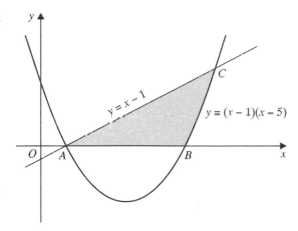

Fig. 1

Figure 1 shows the line $y = x - 1$ meeting the curve with equation $y = (x - 1)(x - 5)$ at A and C. The curve meets the x-axis at A and B.

(a) Write down the coordinates of A and B and find the coordinates of C.

(b) Find the area of the shaded region bounded by the line, the curve and the x-axis.

(11 marks)

7. A geometric series has 4th term 40 and 9th term 1.25.

(a) Find the first term and the common ratio of the series.

(b) Show that the sum S_n of the first n terms is given by

$$S_n = 640 \left[1 - \left(\frac{1}{2} \right)^n \right].$$

A sequence of terms is given by taking $n = 1, 2, \ldots$ in the formula for S_n.

(c) Explain the behaviour of this sequence for large values of n.

(12 marks)

8. A rectangular tank is made of thin sheet metal. The tank has a horizontal square base, of side x cm, and no top. When full the tank holds 500 litres.

(a) Show that the area, A cm^2, of sheet metal needed to make this tank is given by

$$A = x^2 + \frac{2\,000\,000}{x}.$$

(b) Given that x can vary, find the minimum value of A and the value of x for which this occurs.

(c) Prove that this value of A is a minimum.

(12 marks)

9.

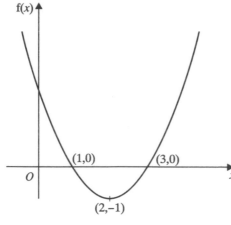

Fig. 2

Figure 2 shows the graph of the quadratic function f whose domain is the set of real numbers. The graph meets the x-axis at $(1,0)$ and $(3,0)$ and the stationary point is $(2,-1)$.

(a) State the range of f and find the equation of the graph in the form $y = f(x)$.

(b) On separate axes, sketch the graphs of
(i) $y = f(x + 2)$, (ii) $y = f(2x)$.

(c) On each graph write in the coordinates of the points at which the graph meets the coordinate axes and write in the coordinates of the stationary point.

(13 marks)

10.

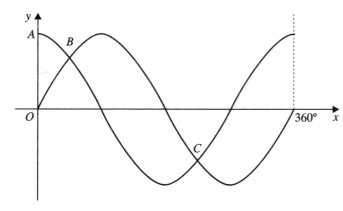

Fig. 3

The diagram shows a sketch of $y = \sin x°$ and $y = \cos x°$ for $0 \leqslant x \leqslant 360$.

(a) Sketch a copy of the diagram and identify which graph is $y = \cos x°$.

(b) Find the coordinates of A, B and C.

(c) Sketch the graph of $y = \cos 2x°$ on the same axes.

(*d*) State how many intersections there are between the curves $y = \sin x°$ and $y = \cos 2x°$ in the interval $0 \leqslant x \leqslant 360$, and hence find the values of x for which $\cos 2x° = \sin x$ in the interval $0 \leqslant x \leqslant 360$.

(*e*) Name the curve in your diagram which coincides with the curve $y = \sin(x° + 90°)$.

(15 marks)

Answers

Exercise 1A

1   $11x^7 + 2x^6 + 5x^5 + 3x^2 + 2x$

2   $3x^5 + 4x^4 - 2x^3 + x^2 + 6x + 2$

3   $2x^3 + 2x^2 - 8$ 4 $6x^4 + 5x^3 - 7x^2 + 10x$

5   $5x^6 + 3x^5 - 2x^4 + x^3 + 7x + 7$

6   $-3x^4 + 8x^3 - x^2 + 9x - 3$

7   $5x^7 - 6x^6 + 5x^4 - 4x^3 - 2x - 5$

8   $-2x^3 + 3x^2 + 10x - 7$

9   $x^5 + 2x^3 + 7x - 9$ 10 $3x^3 - 7x^2 + 7x + 1$

11   $-2x^5 + 9x^3 - 9x^2 + 3x - 1$

12   $-3x^5 + 4x^3 + 9x - 9$

13   $6x^4 - 3x^3 + 10x^2 + 4x + 2$

14   $3x^3 + 5x - 2$ 15 $6x^5 - 3x^3 - 7x + 3$

16   $7x^6 + 3x^3 + 7x - 10$

17   $8x^4 - 6x^3 - 2x^2 + 8x + 15$

18   $12x^5 + 8x^4 + 37x^3 + 32x^2 + 40x + 24$

19   $10x^7 - 26x^5 - 17x^4 - 12x^3 + 3x$

20   $24x^4 + 8x^3 + 20x^2 + 15x - 7$

21   $6x^5 - 9x^4 - x^3 - 6x^2 + 16x - 6$

22   $16x^8 - 28x^7 - 66x^6 + 66x^5 - 6x^4 - 23x^3 - 6x^2$
    $+81x - 42$

23   $2x^7 - 4x^6 + 14x^5 - 13x^4 + 20x^3 - 61x^2 + 63x$
    -18

24   $24x^7 - 16x^6 - 41x^5 + 6x^4 - 9x^3 + 23x^2 + 22x$
    -12

25   $6x^7 - 2x^6 - 29x^5 + 24x^4 - 16x^3 + 4x^2 + 35x$

Exercise 1B

1   $x^2(2 \quad 3x)$ 2 $y(5y^4 + 2y + 7)$ 3 $2x(x + 9)$

4   $8(5y - 1)$ 5 $6x(y + 3)$ 6 $5xy(y^2 + 3x)$

7   $2(x - 3y + 5xy)$ 8 $x(x^2 + 3x - 6)$

9   $3pq(2q + 3p)$ 10 $2c(c + 3d)$

11   $(x - 6)(x + 6)$ 12 $(y - 9)(y + 9)$

13   $(10 - b)(10 + b)$ 14 $2(2a - 3b)(2a + 3b)$

15   $2(3 - 5b)(3 + 5b)$ 16 $a^2(1 - b)(1 + b)$

17   $2(x - 3)(x + 3)$ 18 $3(x - 3)(x + 3)$

19   $17(a - 2)(a + 2)$ 20 $3(4 - 7c)(4 + 7c)$

21   $(x + 3)^2$ 22 $(x - 4)^2$ 23 $(x + 7)^2$

24   $(x - 9)^2$ 25 $3(x - 6)^2$ 26 $(3x + 5)^2$

27   $(2x - 7)^2$ 28 $5(3x - 1)^2$

29   $2(3x^2 + 6x + 1)$ 30 $(5x - 3)^2$

31   $(5x + 2)^2$ 32 $(3x - 7)^2$ 33 $(x + 2)^3$

34   $(x - 3)^3$ 35 $(2x + 1)^3$ 36 $(3x - 1)^3$

37   $(2x + 3)^3$ 38 $(2x - 5)^3$ 39 $y(x + 2)^3$

40   $3(x - 3)^3$

Exercise 1C

1   $(x + 2)(x + 7)$ 2 $(x - 5)(x - 3)$

3   $(x - 4)(x + 3)$ 4 $(x + 12)(x + 1)$

5   $(x - 10)(x + 1)$ 6 $(x - 2)(x - 1)$

7   $(x + 7)(x + 8)$ 8 $(x + 6)(x - 2)$

9   $(x + 5)(x + 2)$ 10 $(x - 8)(x + 5)$

11   $(x - 4)(x - 3)$ 12 $(x - 9)(x - 5)$

13 $(x+9)(x+2)$ **14** $(x-3)(x-7)$

15 $(x+5)(x+4)$ **16** $(x-2)(x-15)$

17 $(x-10)(x-12)$ **18** $(x+15)(x-12)$

19 $(x-11)(x+10)$ **20** $(x+31)(x-20)$

Exercise 1D

1 $(2x+3)(x+2)$ **2** $(2x-3)(x+3)$

3 $(3x+1)(x+2)$ **4** $(5x+1)(x-3)$

5 $(3y+7)(y+2)$ **6** $(2y+1)(y-5)$

7 $(7y-2)(y-4)$ **8** $(5y-2)(y-5)$

9 $(3y+5)(y-5)$ **10** $(3y-2)(y+2)$

11 $(2x+3)(2x+1)$ **12** $(3x-2)(2x+1)$

13 $(4x-3)(2x+5)$ **14** $(5x-4)(2x-3)$

15 $(4x-3)(3x+1)$ **16** $(5x-7)(2x+5)$

17 $(4x-1)(3x-2)$ **18** $(3x+2)(3x+5)$

19 $(2x+7)(2x-3)$ **20** $(7x+3)(5x-2)$

21 $x(x-7)(x+3)$ **22** $x(2-x)(3-2x)$

23 $x^3(5x+2)(2x+1)$ **24** $4(3+5x)(1-2x)$

25 $y(3x-1)(2x+7)$

Exercise 1E

1 $3\sqrt{3}$ **2** $3\sqrt{5}$ **3** $9\sqrt{2}$ **4** $4\sqrt{3}$

5 $5\sqrt{3}$ **6** $7\sqrt{3}$ **7** $9\sqrt{7}$ **8** $4\sqrt{7}$

9 $\sqrt{3}$ **10** $\sqrt{2}$ **11** 3 **12** 3 **13** $17\sqrt{3}$

14 $\sqrt{2}$ **15** $-4\sqrt{5}$ **16** $10\sqrt{6}$ **17** $4\sqrt{7}$

18 $\dfrac{\sqrt{2}}{2}$ **19** $\dfrac{\sqrt{7}}{7}$ **20** $\dfrac{7\sqrt{5}}{5}$ **21** $\dfrac{\sqrt{6}}{9}$

22 $\dfrac{1}{2}$ **23** $\dfrac{1}{3}$ **24** $\dfrac{\sqrt{7}}{7}$ **25** $\dfrac{\sqrt{3}}{6}$

26 $-\sqrt{2}-1$ **27** $\dfrac{\sqrt{3}+1}{2}$ **28** $\dfrac{\sqrt{5}+1}{2}$

29 $\dfrac{\sqrt{7}+4}{3}$ **30** $\dfrac{3+\sqrt{7}}{2}$ **31** $-\dfrac{4+\sqrt{11}}{5}$

32 $\dfrac{\sqrt{7}+\sqrt{3}}{4}$ **33** $\frac{1}{3}(\sqrt{11}+\sqrt{5})$

34 $\frac{7}{10}(\sqrt{13}+\sqrt{3})$ **35** $\frac{7}{3}(\sqrt{7}-2)$

36 $\dfrac{4\sqrt{3}-8-2\sqrt{11}+\sqrt{33}}{5}$ **37** $28+7\sqrt{14}$

38 $4-\sqrt{15}$ **39** $\frac{1}{3}(10-\sqrt{91})$

40 $\frac{1}{10}(33-\sqrt{989})$

Exercise 2A

1 6 **2** -1.8 **3** 2 **4** 3 **5** 1 **6** 2

7 30 **8** $\frac{5}{6}$ **9** $-\frac{1}{4}$ **10** 7 **11** -23

12 6 **13** $(7,4)$ **14** $(3,-1)$ **15** $(-2,3)$

16 $(-4,-5)$ **17** $(-\frac{1}{2},1\frac{1}{2})$ **18** $(1,-2)$

19 $(\frac{1}{2},\frac{1}{2})$ **20** $(-3,-2)$ **21** $(-2,+6)$

22 $(-14,-2)$ **23** $(2,-3)$ **24** $(2,\frac{1}{2})$

25 $(-3,-9)$ **26** $(-1,-2\frac{1}{2})$ **27** $(-\frac{2}{3},\frac{3}{4})$

Exercise 2B

1 ±7 **2** ±2.5 **3** ±12 **4** $0,-5$

5 $0,7$ **6** $2,-1$ **7** $-3,2$ **8** $-2,-5$

9 $4,5$ **10** $-3,-3$ **11** $-4,-8$

12 $10,-1$ **13** $4,-3$ **14** $-6,2$ **15** $12,-1$

16 $-1,-12$ **17** $2,-\frac{1}{2}$ **18** $1,-\frac{1}{2}$

19 $-2,-\frac{1}{2}$ **20** $2,\frac{1}{3}$ **21** $\frac{2}{3},-7$ **22** $5,-\frac{2}{3}$

23 $\frac{1}{4},\frac{1}{3}$ **24** $\frac{3}{2},-\frac{2}{3}$ **25** $\frac{5}{4},\frac{3}{2}$ **26** $\frac{9}{4},-\frac{3}{7}$

27 $-\frac{2}{3},-\frac{4}{5}$ **28** $\frac{9}{5},-\frac{1}{4}$ **29** $\frac{3}{10},-\frac{7}{11}$

30 $\frac{22}{37},-\frac{3}{5}$ **31** $2.41,-0.41$ **32** $-7.16,-0.84$

33 $4.19,-1.19$ **34** $0.66,-0.46$

35 $-0.23,-0.63$ **36** $1.18,-0.43$

37 $1.43,0.23$ **38** $0.53,-1.13$

39 $1.72,-0.39$ **40** $2.65,0.85$

Exercise 2C

1 $(2,-1),(-\frac{32}{17},\frac{49}{17})$ **2** $(0,1),(\frac{4}{5},-\frac{3}{5})$

3 $(-2,0),(\frac{1}{3},\frac{7}{3})$ **4** $(1,-3),(6,-\frac{1}{2})$

5 $(6,3),(4\frac{1}{2},4\frac{1}{2})$ **6** $(2,1),(-2,-1)$

7 $(3,12),(1,4)$ **8** $(8,+5),(-5,-8)$

9 $(1,-2),(-2\frac{3}{7},-\frac{2}{7})$ **10** $(-1,3),(-2,2)$

Exercise 2D

1 $x > 1$ 2 $x < 3$ 3 $x < -2$ 4 $x < -2$
5 $x \geqslant 2$ 6 $x \leqslant 3$ 7 $x < 6\frac{1}{2}$
8 $x \leqslant -2\frac{1}{3}$ 9 $x < \frac{1}{4}$ 10 $x \geqslant 3$
11 $x > 3$ 12 $x > 5$ 13 $x \geqslant 4$ 14 $x < 7$
15 $x \leqslant 5$ 16 $x \geqslant 3\frac{1}{2}$ 17 $x < -3$
18 $x \leqslant -\frac{29}{7}$ 19 $x < \frac{5}{3}$ 20 $x \geqslant \frac{5}{8}$
21 $x \geqslant -2, x \leqslant -5$ 22 $2 \leqslant x \leqslant 3$
23 $x > 3, x < -4$ 24 $3 < x < 6$
25 $-2 \leqslant x \leqslant 9$ 26 $x < -7, x > -4$
27 $x < \frac{3}{2}, x > 4$ 28 $x < -2, x > \frac{4}{3}$
29 $-\frac{2}{3} < x < 7$ 30 $x < 3, x > \frac{7}{2}$
31 $-4 \leqslant x \leqslant -\frac{5}{2}$ 32 $x < -1, x > \frac{9}{5}$
33 $x \leqslant -9, x \geqslant -\frac{7}{3}$ 34 $\frac{3}{4} < x < 5$
35 $x < \frac{5}{3}, x > 6$ 36 $-\frac{2}{3} \leqslant x \leqslant 2$
37 $-2 < x < \frac{3}{2}$ 38 $-\frac{1}{4} \leqslant x \leqslant \frac{2}{3}$
39 $x < -\frac{2}{3}, x > \frac{3}{7}$ 40 $-\frac{4}{3} < x < \frac{5}{2}$

Exercise 3A

1 $-3 \leqslant y \leqslant 9$ 2 $2 \leqslant y \leqslant 6$ 3 $0 \leqslant y \leqslant 16$
4 $\frac{1}{10} \leqslant y \leqslant \frac{1}{2}$ 5 $\frac{1}{4} \leqslant y \leqslant 16$
6 (a) $\frac{1}{5}$ (b) $\frac{1}{2}$ (c) 1 7 $\frac{1}{8} \leqslant y \leqslant 1$
8 $-1 \leqslant x \leqslant \frac{8}{3}$ 9 $-3 \leqslant x \leqslant 6$
10 $0 \leqslant x \leqslant 6$

Exercise 3B

1 (a) (i) $fg : x \mapsto 9 - 8x, x \in \mathbb{R}$
    (ii) $gf : x \mapsto 27 - 8x, x \in \mathbb{R}$
    (iii) $fh : x \mapsto 13 - 6x, x \in \mathbb{R}$
    (iv) $hf : x \mapsto 18 - 6x, x \in \mathbb{R}$
    (v) $gh : x \mapsto 12x - 13, x \in \mathbb{R}$
    (vi) $hg : x \mapsto 12x - 6, x \in \mathbb{R}$
  (b) (i) -31 (ii) -15 (iii) 13 (iv) -12
    (v) -39 (vi) -1
  (c) (i) 3 (ii) 2 (iii) -1
2 (a) $1\frac{1}{2}$ (b) $\frac{5}{6}$ (c) $2\frac{2}{3}$ $x = -\frac{1}{4}$ or 4

3 (a) 0 (b) 8 (c) 35 (d) 3
4 $p(x) = x^3$, $q(x) = x + 2 \rightarrow qp = f$
  $p(x) = \frac{1}{x}$, $q(x) = x - 3 \rightarrow qp = g$
  $fg(2) = -13\frac{5}{8}$ $gf(2) = -2\frac{9}{10}$
5 $a \mid b = 1$
6 (a) $1 + \frac{x^2}{2}$ (b) $\left(1 + \frac{x}{2}\right)^2$ (c) $\frac{x + 6}{4}$
8 $fg(x) = 18x^2 - 24x + 9$
  $gf(x) = 6x^2 + 1$
  $ff(x) = 8x^4 + 8x^2 + 3$
  $gg(x) = 9x - 8$

Exercise 3C

1 $f^{-1}(x) = \frac{1}{2}x$ 2 $f^{-1}(x) = \frac{x - 3}{2}$
3 $f^{-1}(x) = \frac{1 - x}{2}$ 4 $f^{-1}(x) = -x^{\frac{1}{2}}$
5 $f^{-1}(x) = \frac{1}{x}$ ($f(x) = \frac{1}{x}$ is a self-inverse function)
6 $x - p$ 7 $\frac{x}{p}$ 8 $p - x$ (self-inverse)
9 $\frac{x - q}{p}$
10 $\frac{p}{x}$ $(x \neq 0)$ (self-inverse)
11 $\frac{1 - qx}{px}$ $(x \neq 0)$
12 $y \in \mathbb{R}^+; \frac{1}{x^{\frac{1}{2}}}$ 13 $0 < y < 1; \frac{1 - x}{x}$
14 $y \in \mathbb{R}^+; (x + 1)^{\frac{1}{2}} - 1$ 15 $y > 5; (x - 1)^{\frac{1}{2}} - 2$
16 $A = 5, B = 4; k = 5, y \geqslant 4;$
  $f^{-1}(x) = (x - 4)^{\frac{1}{2}} + 5, x \geqslant 4$
17 $g^{-1}(x) = \frac{1}{x} + 4, x \neq 0$
18 Domain of inverse function $\frac{1}{4} \leqslant x \leqslant 8$

Exercise 3D

3 (a) $y = x^2, y = x^4$ (b) $y = x^3$
4 $\left(\frac{1}{2}, \frac{1}{2}\right)$ and $(1,1)$

Review exercise 1

1. (a) $x(x-7)$ (b) $(x-9)(x+2)$
   (c) $(x-3-y)(x-3+y)$
2. (a) $-\frac{1}{2}, -2\frac{1}{2}$ (b) $1.2, -4.7$
3. $2x^3 + 9x^2 + x - 12$ 4. $(-\frac{1}{2}, 0), (-1, -1)$
5. $0.62, -1.62$ 6. $(1, -2)$ (7) $x + 7$
8. -1 9. 2 10. (a) $0.78, -1.28$ (b) -2
11. $\{2, 3, 4\}$ 12. (a) 11 (b) $p - 7q - 1$
13. (a) $f^{-1} : x \mapsto \dfrac{x+1}{6}, x \in R$
    (b) $fg : x \mapsto \dfrac{25-x}{x-1}, x \in R, x \neq 1$ (c) $-\frac{1}{3}, \frac{3}{2}$
14. $(2, 3)$ 15. $(7, 3), (-7, -4)$
16. $\dfrac{2\sqrt{7} - 5}{3}$ 17. (a) $1, 4$ (b) $5.37, -0.37$
18. (a) $5x(x-4)$ (b) $(a-6)(a+4)$
    (c) $(3x-1)(3x+1)$ 19. $-1 < x < \frac{1}{2}$
20. $2x + 6y - 9z$ 21. $m = 9, n = 2$; range $y \leqslant 9$
22. (a) (i) $(2x-y)(2x+y)$
    (ii) $(2x-5)(x-2)$ (iii) $2x(x^2+3)$
    (b) $a = 4, b = -3$
23. $(7, -3), (5, -\frac{5}{3})$ 24. $-\dfrac{\sqrt{3}}{3}$
25. (a) (i) $(x-7)(x+5)$ (ii) $4(x-2)^2$;
    (b) $(a-b)(a+b), 4\,960\,000$
26. (a) $0, -4$ (b) $2, -6$ (c) $-0.21, -4.79$
27. $-1 < x < 4$
28. (a) $2.54, -3.54$ (b) $x^3 - 3x - 2$
29. (a) $\dfrac{1}{1-x}$ (b) x (c) $\dfrac{1}{1-x}$
30. $2\frac{1}{3}, 3$ 31. $3, -4$ 32. $41 - 17\sqrt{7}$
33. (a) $2(p+3)(p-q)$ (b) $0.88, -1.88$
34. $x = 14$ 35. $-5 < x < 1\frac{3}{4}$
36. (a) $8 - 2x^2 + 2x^3$ (b) 4
37. (a) $2 - 5x + 2x^2 + x^3$ (b) $1, -17$
38. (a) $f(x) \geqslant -1$
    (b) $fg : x \mapsto 2x^2 + 12x + 17, x \in \mathbb{R}$
39. $(5, -2)$ 40. $(2, 1)$ 41. $\dfrac{66 - 15\sqrt{11}}{19}$
42. (a) $(4p-q)(4p+q)$ (b) $(x-8)(x+3)$

(c) $(3a-2)(b+3)$ 43. $-2.38, -4.62$
44. $-\frac{1}{2} < x < 3$ 45. (a) $1.7, -0.34$
    (b) $-12 + 14x + 19x^2 - 15x^3$
46. (a) $fg : x \mapsto \dfrac{x+2}{3x}, x \neq 0$
    (b) $g^{-1} : x \mapsto \dfrac{4+2x}{1-x}, x \neq 1$
47. $a = 2, b = -1$ 48. $(\frac{1}{3}, 12), (8, \frac{1}{2})$
49. $11 - 5\sqrt{3}$ 50. (a) (i) $(x-13)(x+1)$
    (ii) $q(2p-5q)(2p+5q)$
    (iii) $(c-2d)(2a-b)$ (b) $(-\frac{1}{2}, -4)$
51. (a) $3(x-1)(x+1), \dfrac{x-4}{3(x+1)}$
    (b) $3.08, -1.08$ 52. $x \leqslant 5$
53. (a) $14x + y - 11z$ (b) 4
54. (a) $y \in \mathbb{R}, y \neq 1$ (b) $ff(x) = x$
    (c) $f^{-1}(x) = \dfrac{x+3}{x-1}, x \neq 1$
55. (a) $(1, -3)$ (b) -7 56. $(3, 1), (-5, -7)$
57. $\dfrac{\sqrt{17} + \sqrt{13}}{4}$
58. (a) $2(x-3y)(x+3y)$ (b) $(3a-2b)(c-2d)$
59. $2, -\frac{3}{2}$ 60. $x > 4$ or $x < -3$
61. (a) $1.90, -1.23$ (b) $-13, -43, 27$
62. (a) $ff : x \mapsto \dfrac{10x+1}{x+5}, x \in \mathbb{R}, x \neq -5$
    (b) $f^{-1} : x \mapsto \dfrac{2x+1}{x-3}, x \in \mathbb{R}, x \neq 3$
63. $(3, -1)$ 64. $(\frac{1}{5}, \frac{13}{10}), (\frac{1}{3}, \frac{3}{2})$ 65. $\dfrac{7\sqrt{5} - 32}{19}$
66. (a) $3p(p-3)$ (b) $(y+1)(y+9)$
67. (a) $3.89, -0.39$ (b) 14
68. (a) $0.719, 2.78$ (b) $-6x^3 + 13x^2 - 9x + 2$
69. (a) $\frac{1}{5}$ (b) $g^{-1} : x \mapsto \dfrac{1}{x}, x \in \mathbb{R}, x \neq 0$
    (c) $2, -1$ 70. $(2, -1)$
71. $(1, 0), (\frac{3}{5}, \frac{4}{5})$
72. (a) $(4x-3y)(4x+3y)$ (b) $(3a-1)(b-3)$
    (c) $(y+2)(3y-5)$ 73. $1.42, -0.422$
74. $9, 4, -16$; $x \leqslant -5\frac{1}{3}$ $x \geqslant -2\frac{2}{3}$
75. (a) $x \leqslant 14$ (b) $-4 < x < 4$

76 (a) $2, -\frac{2}{3}$ (b) $2.58, -0.58$

77 (b) $(gf)^{-1} : x \mapsto \dfrac{2x}{x-2}, x \in \mathbb{R}, x \neq 2$

78 (a) 8 (b) ± 4 (c) $4a + 3b + c$

79 (a) $2, 6$ (b) ± 5 (c) $4.11, -0.61$

80 (a) $0, -3$ (b) $-\frac{1}{2}, \frac{5}{6}$ (c) $-0.63, -2.37$

81 (a) $(3x - 5y)(3x + 5y)$ (b) $(p + 2q)(r - 3s)$
(c) $(3a + 2)(a - 1)$

82 $-\frac{1}{2} \pm \sqrt{2}$ **83** $-1 \leqslant p \leqslant 9$

84 (a) 22 (b) $x^3 + x^2 + x - 3$ (c) $0, 2$

85 (a) $1 \leqslant y < 25$
(b) $f^{-1} : x \mapsto \dfrac{2x + 25}{3x}, x \in \mathbb{R}, 1 \leqslant x < 25$
(c) $fg : x \mapsto \dfrac{25}{3x^2 - 2}, x \in \mathbb{R}, 1 < x \leqslant 3$
(d) $\frac{7}{6}, 3$

86 $\frac{3}{2}, \frac{5}{2}$ **87** $(0, 1), (-2, -3)$

88 $\dfrac{\sqrt{5}(\sqrt{19} - \sqrt{3})}{80}$

Exercise 4A

1 $\sqrt{10} = 3.16$ (3 s.f.) **2** $\sqrt{34} = 5.83$ (3 s.f.)

3 $\sqrt{122} = 11.0$ (3 s.f.) **4** $5\sqrt{5} = 11.2$ (3 s.f.)

5 13 **6** $\sqrt{13} = 3.61$ (3 s.f.)

7 $\sqrt{52} = 7.21$ (3 s.f.) **8** 12

9 $\sqrt{212} = 14.6$ (3 s.f.) **10** $\sqrt{29} = 5.39$ (3 s.f.)

11 $5\sqrt{10} = 15.8$ (3 s.f.) **12** $\sqrt{13} = 3.61$ (3 s.f.)

15 $PQ = 6.32, QR = 5.39, PR = 4.12$

16 (a) 14.4 (b) 10.2

20 $a = 3, b = 4$

Exercise 4B

1 2 **2** -1 **3** 3 **4** $-\frac{8}{5}$ **5** 3

6 -3 **7** $7\frac{1}{2}$ **8** 2 **9** $\frac{13}{5}$ **10** 3

11 $-\frac{17}{9}$ **12** $-\frac{1}{5}$ **13** 3 **14** -1

15 $AB \; \frac{8}{5}, \; BC \; -\frac{4}{9}, \; CA \; -3$

16 $AB \; \frac{2}{3}, \; BC \; 2, \; CA \; -2$

17 $AB \; -\frac{7}{2}, \; BC \; -1, \; CA \; -\frac{4}{9}$

18 both have gradient $-\frac{1}{2}$, so AB and C lie on a straight line

19 both have gradient $\frac{3}{2}$ and are on same line

Exercise 4C

1 (a) 5 (b) -7 (c) -2 (d) -1
(e) 1.5 (f) $\frac{7}{3}$ (g) $\frac{2}{3}$ (h) $\frac{3}{4}$
(i) $-\frac{3}{2}$ (j) $-\frac{4}{3}$ (k) $\frac{5}{2}$ (l) $-\frac{6}{5}$

2 (a) -4 (b) 6 (c) 3 (d) -2
(e) $\frac{7}{2}$ (f) $\frac{5}{3}$ (g) 3 (h) $\frac{7}{5}$
(i) $\frac{7}{2}$ (j) $\frac{8}{5}$ (k) $-\frac{1}{2}$ (l) -4

3 (a) $(0, 4)$ (b) $(0, 7)$ (c) $(0, \frac{3}{2})$
(d) $(0, -\frac{2}{3})$ (e) $(0, \frac{5}{2})$ (f) $(0, 3)$
(g) $(0, \frac{7}{3})$ (h) $(0, \frac{3}{2})$

4 $y = 2x + 3$ **5** $y = 5x - 4$

6 $y = 2 - x$ **7** $y = -\frac{2}{3}x + 1$

8 $y = -\frac{1}{2}x - 7$ **9** $y - 1 = 2(x - 5)$

10 $y - 3 = -3(x - 8)$ **11** $y - 5 = +\frac{1}{2}(x - 5)$

12 $y + 2 = -\frac{4}{5}(x - 3)$ **13** $y - 6 = 7(x - 3)$

14 $y + \frac{1}{2} = 3(x - 1)$

15 (a) $2x - 3y - 1 = 0$ (b) $y = 3x - 2$
(c) $y + x - 1 = 0$ (d) $x + 3y - 11 = 0$
(e) $13x + y + 20 = 0$ (f) $6x - y - 26 = 0$
(g) $2y - 3x + 18 = 0$ (h) $2x + y + 7 = 0$

Exercise 5A

1 x^4 **2** a^6 **3** $4a^3$ **4** $30a^2b^2$

5 x^{10} **6** y^{12} **7** $12y^7$ **8** $24y^8$

9 $60a^5b^4$ **10** $36x^7y^4$ **11** $96p^{10}q^4$

12 $48a^5b^5$ **13** $30p^5q^9$ **14** $30x^3y^2z^3$

15 $48a^4b^3c^7$ **16** $72a^3b^2$ **17** $360a^5b^7c^6$

18 $84p^4q^4$ **19** $6a^4b^4pq^3$ **20** $\frac{1}{10}a^6p^4q^3$

Exercise 5B

1 a^{30} **2** x^{14} **3** a^8b^{12} **4** $p^{15}q^{20}$

5   x^4 6 a^4 7 p^4 8 $5p^3$

9    $3a^2$ 10 $12a^9$ 11 $2a^8$ 12 $4p^5$

13    $4x^4$ 14 $5x^7$ 15 $3ab^4$ 16 $2a^2b$

17    $5a^5b$ 18 $36a^4b^4c^2$ 19 $6a^2bc^2$

20    $3x^2y^3z$

Exercise 5C

1   2   2   4   3   2   4   3   5   27

6   $\frac{1}{4}$ 7   1   8 $\frac{1}{100}$ 9 $-\frac{1}{343}$

10   $\frac{5}{2}$ 11 $1\frac{1}{3}$ 12 $\frac{3}{2}$ 13 $\frac{4}{9}$ 14   9

15   1   16 $\frac{6}{5}$ 17 $1\frac{11}{25}$ 18 $\frac{25}{4}$

19   $\frac{5}{4}$ 20 $\frac{13}{9}$

Exercise 5D

1   (a)   3.2   (b)   1.2   (c)   9.8   (d)   0.57

2   (a)   24   (b)   0.64   (c)   4.2

3   (a)   0.74   (b)   6.7   (c)   15

    (d)   1.9   (e)   2.8

Exercise 5E

1   (a) -0.92 (b) -0.22 (c)   0.53

    (d)   1.19   (e)   1.55   (f)   0.37

    (g)   0.55   (h)   6.05

2   (b) and (c) are translations of (a) in the direction of Oy.

Exercise 6A

1   (a)   3, 6, 9, 12   (b)   1, 6, 11, 16

    (c) -1, 5, 15, 29   (d)   1, 2, 4, 8

2   (a)   11, 13; $2n + 1$ (b)   243, 729; 3^n

    (c) $\frac{5}{6}, \frac{6}{7}; \dfrac{n}{n+1}$ (d)   65, 129; $1 + 2^{n+1}$

    (e)   30, 42; $n^2 + n$

3   (a)   17, 26   (b)   41, 122   (c)   677, 458 330

4   1, $\frac{1}{4}$, $\frac{1}{9}$, $\frac{1}{16}$: converges to zero

5   (a)   oscillates between $-\frac{1}{2}$ and $\frac{1}{2}$

    (b)   periodic with period 8

    (c)   converges to zero

6   0, $\frac{1}{3}$, $\frac{1}{2}$, $\frac{3}{5}$; the nth term approaches 1 as n approaches ∞

7   (a) converges to 3; (c) and (d) both converge to zero; (b) and (e) are non-convergent

8   (a)   15, 22 : $(7n + 1)$

    (b) -13, -19 : $(-6n + 5)$

    (c) -3, -19 : $(13 - 8n)$

9   $30°$, $29.7°$, $29.403°$, $29.108\,97°$

    20th term $24.785°$, 100th term $11.092°$

Exercise 6B

1   3, 2, 0, -4 . . . : divergent

2   3, 10, 24, 52 . . . : divergent

3   0, 1, $\frac{3}{2}$, $\frac{7}{4}$ . . . : converges to 2

4   3, 2, $\frac{5}{3}$, $\frac{14}{9}$ . . . : converges to $1\frac{1}{2}$

5   2, $\frac{1}{5}$, $-\frac{4}{25}$, $-\frac{29}{125}$ . . . : converges to $-\frac{1}{4}$

6   $\frac{3}{2}$, $\frac{7}{5}$, $\frac{17}{12}$ . . . : limiting value $\sqrt{2}$

7   33, 63, 153

8   149, 2389, 19 115

9   (a)   diverges   (b)   converges to zero

    (c)   oscillates   (d)   converges to zero

10   -6

11   limiting value 1 in both cases

12   $u_n = u_{n-1}u_{n-2}$, $u_1 = 1$, $u_2 = 3$

    10th term $= 3^{34}$, 13th term $= 3^{144}$

Exercise 6C

1   $3 + 6 + 9 + 12 + \ldots + 3n$

2   $1 + 3 + 5 + 7 + \ldots + 2n - 1$

3   $1 + 8 + 27 + 64 + \ldots + n^3$

4   $2 + 4 + 8 + 16 + \ldots + 2^n$

5   $1 + \frac{2}{3} + \frac{3}{5} + \frac{4}{7} + \ldots + \frac{n}{2n-1}$

6   $1 + 4 + 7 + 10$ 7 $6 + 12 + 20 + 30$

8   $1 + 2 + 5 + 12$ 9 $1 + \frac{1}{2} + \frac{1}{3} + \frac{1}{4}$

10   $13 + 16 + 19 + 22$

11   $3 + 5 + 7 + 9 + 11 + 13$

12   $9 + 16 + 25 + 36 + 49$

13   $24 + 48 + 96 + 192 + 384$

14   $293 + 296 + 299 + 302 + 305 + 308 + 311$

15   $\frac{13}{15} + \frac{15}{17} + \frac{17}{19} + \frac{19}{21}$

Exercise 6D

1   $3, 425$ 2 $4, 777$ 3 $-7, -536$

4   $0.25, 116.25$ 5   £25, £56 925   6   276

7   432   8   196   9   9900   10   37 500

11   $7, 1530$ 12 $314, 4940$ 13 $10, -1, 55$

14   31 terms, 50 terms

15   (a) £7600   (b) £30 200

16   $1, 55, (6n - 5)$ 17 $-2.5, \frac{1}{2}, 735$

18   $x = 3; -60$ 19 $\frac{r-4}{6}, 16\frac{1}{2}$ 20   52

21   9   22   7   23 $-1, 9$ 24   15 000

25   $\frac{y - x}{2(n - 2)}$

Exercise 6E

1   (a) $5120, 10\,230$ (b) $-5120, -3410$

    (c) $\frac{5}{256}, 19\frac{251}{256}$

2   (a) $12\,285$ (b) -4095 (c) 8.93

    (d) 1.79

3   (a) $\frac{1}{32}, 7\frac{31}{32}$ (b) $-2048, -4092$

    (c) $-\frac{16}{729}, 36.0$ (d) $-28.24, -745.8$

4   series (b) has no sum to infinity

    (a) 8   (c) 36   (d) -1000

5   $\frac{1}{9} - \frac{1}{3} + 1 - \ldots; -14\,762.\dot{2}$

6   (a) $\frac{1}{3}[1 - (\frac{1}{10})^n]; \frac{1}{3}$

    (b) $\frac{32}{3}[1 - (-\frac{1}{2})^n]; \frac{32}{3}$

    (c) $\frac{16}{5}[1 - (-\frac{3}{2})^n]$; diverges

7   $-\frac{2}{5}, 5.00$ 8 $3, \frac{2}{3}, 8.93$

9   $2, -3$; no sum to infinity because $|r| > 1$

10   36 years

11   (a) 36   (b) 40

12   $\frac{64}{27}, \frac{2}{3}; 81$

14   (a) £55 300   (b) £20 441 700

14   (a) 105   (b) 11.1

15   -14.1

Exercise 6F

1   $p = 18, q = 2; 27$ 2 $15\frac{15}{16}$ 3   90 m

4   $7, 8$ 5 $243\sqrt{6}; 364(\sqrt{6} + \sqrt{2})$

6   $6, 2$ 7 $-\frac{8}{9}$

8   (a) -10.1 (b) 49

9   $\dfrac{10^8}{9}, -\dfrac{10^8}{11}$

10   76.21   11 $8(4n + 3), n > 1$

Review exercise 2

1   $4x + 3y = 29$

2   (a) 5   (b) $\frac{1}{9}$ (c) $\frac{1}{3}$ (d) 1

3   $1.5, 6.25$

5   (a) $\frac{2}{3}$ (b) $\frac{1}{16}$ (c) 64   (d) 1

6   22

7   (a) $11, -4$ (b) 25.5

8   (a) $\frac{1}{48}, \frac{1}{96}$ (b) $\frac{5}{8}, \frac{6}{243}$ (c) $120, 720$

9   (a) 4   (b) 64   (c) $\frac{1}{2}$

10   (a) 1   (b) $\frac{2}{3}$ (c) 54

11   $\frac{3}{5}$

12   (a) 1   (b) 2   (c) 1   (d) 4

13   $\frac{1}{4}$

14   (a) 14.9   (b) 689   16 $2 \pm \sqrt{2}$

17   (b) £3216   18 $4, -1, 7, 5, 19, 29, 67$

19   (b) $(1, 2)$

20   (a) $\frac{1}{81}$ (b) 3   (c) 32   (d) 0

21   (i) $\frac{10\,100}{3}$ (ii) $\frac{7}{9}, \frac{k}{9}$ 22 $\frac{1}{6}$ units^2

23   $\frac{1}{2}, \frac{3}{2}$

24 (a) (i) £3370 (ii) £5679

(b) $2000[1.11 + 1.11^2 + \ldots + 1.11^{12}]$
$= £50\,423$

25 0, 2

26 (a) $18\frac{1}{2}$ (b) 27.7 (3 s.f.) (c) 32

27 (a) 2 (b) $\frac{1}{10\,000}$ (c) 1 (d) 2

28 (a) (1, 3) (b) (−4, −2) (c) $5\sqrt{2}$

29 $\frac{1}{3}$ **30** (i) 5, −30; 13 (ii) 49.8

31 2, 5, 3.5, 4.25, 3.875, 4.0625, 3.96875

33 32

34 (i) (a) −1, 1, 3 (b) 2400 (c) 25

(ii) (a) 12, −8; 3, $\frac{1}{2}$; (b) −32

35 $5\sqrt{10}$ **36** −2

37 (i) (a) $3\frac{1}{2}$ (b) $-\frac{1}{2}, 5\frac{1}{2}, 11\frac{1}{2}, 17\frac{1}{2}$

(ii) (a) $-\frac{1}{2}, 8$

(b) $-4\frac{1}{2}, 1\frac{1}{2}, -\frac{1}{2}, \frac{1}{6}$; 4, 10, 25, 62.5

(c) $-\frac{1}{3}, \frac{5}{2}$ (d) $-3\frac{3}{8}$

38 5 units^2

39 (i) (a) 3 (b) $\frac{1}{9}$ (c) 1

(ii) −1

40 (i) 4

(ii) (a) 97 (c) 2 (d) $\frac{5}{4}$ (e) 4110

41 1.9, 0.11 **42** $\dfrac{6a^2}{b^2c^2}$

43 (a) 120 (b) 3836.25 (c) 427.5

(d) 4192.5

44 5.5 **45** (a) 24 (b) 8 (c) $\frac{1}{2}$

46 (a) −11 (b) $-\frac{1}{243}$ **47** (3, −2)

48 (a) (i) x (ii) x^8 (iii) x^{10} (b) 16

49 (i) (a) $\frac{5}{12}$ (b) $3\frac{7}{12}$ (ii) (a) $\pm\frac{3}{2}$

(b) 33, −6.6

50 (a) $(1 \times 2) + (2 \times 3) + (3 \times 4) + (4 \times 5) + (5 \times 6) = 70$

(b) $\dfrac{1}{2 \times 1} + \dfrac{1}{3 \times 2} + \dfrac{1}{4 \times 3} + \dfrac{1}{5 \times 4} = \dfrac{4}{5}$

51 (a) $9x + 13y + 19 = 0$

52 −1

53 (i) (a) 7, $1\frac{1}{2}$ (b) 462

(ii) (a) $\frac{1}{2}$ (b) 3 (c) 364.5

54 (3, 11)

55 (i) −3, −1, 1, 3; $n^2 - 4n$; 15

(ii) −5; −4, $\frac{1}{2}$; −8

56 (a) 12 (b) $\frac{1}{27}$ (c) 1

57 (i) (a) $1\frac{1}{2}$ (b) −3 (c) 21 (ii) $\frac{1}{4}$

58 −1, 1, −5, 1, −29, −23, −197

Exercise 7A

1 (a) 180° (b) 90° (c) 30° (d) 45°

(e) 270° (f) 135° (g) 720° (h) 540°

(i) 225° (j) 150°

2 (a) $\dfrac{\pi}{8}$ (b) $\dfrac{\pi}{12}$ (c) π (d) $\dfrac{7\pi}{6}$

(e) $\dfrac{2\pi}{3}$ (f) $\dfrac{3\pi}{4}$ (g) $\dfrac{5\pi}{4}$ (h) $\dfrac{11\pi}{6}$

(i) $\dfrac{5\pi}{3}$ (j) $\dfrac{7\pi}{4}$

3 (a) 57° (b) 138° (c) 286° (d) 201°

(e) 23° (f) 97° (g) 246° (h) 315°

(i) 229° (j) 344°

4 (a) 0.35^c (b) 0.87^c (c) 1.22^c

(d) 2.27^c (e) 2.97^c (f) 4.01^c

(g) 4.36^c (h) 1.48^c (i) 0.66^c

(j) 2.65^c

5 (a) 3 cm, 4.5 cm^2 (b) 14 cm, 49 cm^2

(c) 12 cm, 48 cm^2 (d) 28.5 cm, 135.4 cm^2

(e) 2.75 cm, 7.56 cm^2

(f) 25.2 cm, 113.4 cm^2

(g) 49.5 cm, 272.25 cm^2

(h) 47.5 cm, 225.6 cm^2

(i) 52.64 cm, 294.8 cm^2

(j) 50.63 cm, 210.1 cm^2

6 2.5^c **7** 3.6 cm **8** 1.25^c **9** 8.45 cm

10 (a) 3 cm^2 (b) 0.9375 cm^2

(c) 2.1875 cm^2 (d) 5.6875 cm^2

(e) 6.336 cm^2

Exercise 7B

1. (a) 0.471   (b) 0.451   (c) 1.56
   (d) 0.854   (e) 1.13   (f) 0.271
   (g) 0.441   (h) 0.714   (i) 0.220
   (j) 0.946

2. (a) 0.230   (b) 0.942   (c) 1.53
   (d) 0.842   (e) 0.992   (f) 0.325
   (g) 2.08   (h) 0.842   (i) 0.452
   (j) 0.316

3. (a) 20.0°   (b) 83.6°   (c) 56.7°
   (d) 70.7°   (e) 61.9°   (f) 7.55°
   (g) 33.5°   (h) 51.4°   (i) 28.0°
   (j) 52.0°

4. (a) 0.244   (b) 1.11   (c) 0.728
   (d) 1.07   (e) 0.882   (f) 0.847
   (g) 0.916   (h) 0.997   (i) 1.13
   (j) 1.33

5. (a) 0.962   (b) 0.703   (c) 1.34
   (d) 0.671   (e) 0.701   (f) 0.823
   (g) 0.978   (h) 0.456   (i) 0.355
   (j) 0.752

Exercise 7C

1. (a) 6.79 cm   (b) 6.37 cm   (c) 5.89 cm
   (d) 9.80 cm   (e) 19.7°   (f) 70.6°
   (g) 18.6°   (h) 15.5 cm   (i) 29.3 cm
   (j) 21.3 cm   (k) 8.02 cm   (l) 5.07 cm
   (m) 40.7 cm   (n) 13.4 cm   (o) 31.0 cm
   (p) 13.0 cm   (q) 6.78 cm   (r) 44.5°

2. (a) 4.95 cm   (b) 38.6 cm²

3. (a) 17.6 m   (b) 35.6 m

4. (a) 2.82 cm   (b) 5.30 cm   (c) 43.2°

5. (a) 51.3°   (b) 12.8 m   (c) 23.0 m
   (d) 350 m²

6. (a) 7.5 cm   (b) 6.73 cm   (c) 14.0 cm
   (d) 31.2 cm   (e) 38.2 cm²

7. 1880 cm²

8. (a) 5.79 cm   (b) 8.34 cm   (c) 3.06 cm

   (d) 98.6 cm²

9. (a) 33.0 m   (b) 12.0 m

10. (a) (i) 2.94 km   (ii) 4.05 km
    (b) (i) 8.29 km   (ii) 1.90 km
    (c) (i) 283°   (ii) 8.51 km

11. (a) 9.58 km north, 7.22 km east
    (b) 19.15 km south, 16.07 km east
    (c) 25.18 km
    (d) 112°

Exercise 7D

1. (a) $-\sin 80°$ (b) $\cos 40°$ (c) $\tan 5°$
   (d) $-\tan 5°$ (e) $-\cos 43°$ (f) $\sin 54°$
   (g) $\sin 14°$ (h) $\cos 24°$ (i) $\tan 36°$
   (j) $-\tan 32°$ (k) $-\sin 83°$ (l) $-\cos 68°$
   (m) $-\sin 29°$ (n) $\cos 67°$ (o) $\tan 84°$
   (p) $\tan 36°$ (q) $-\sin 21°$ (r) $-\cos 59°$
   (s) $-\tan 53°$ (t) $\sin 18°$

2. (a) $\frac{\sqrt{3}}{2}$ (b) $-\frac{\sqrt{3}}{2}$ (c) 1 (d) $\frac{1}{2}$
   (e) $-\frac{1}{2}$ (f) $-\frac{1}{2}$ (g) $-\frac{\sqrt{3}}{2}$ (h) $\frac{\sqrt{3}}{2}$
   (i) $\frac{1}{\sqrt{2}}$ (j) $-\frac{1}{\sqrt{2}}$ (k) $\frac{1}{\sqrt{3}}$ (l) $-\frac{1}{\sqrt{3}}$
   (m) $-\frac{1}{2}$ (n) $-\frac{\sqrt{3}}{2}$ (o) $\sqrt{3}$ (p) $\frac{1}{\sqrt{3}}$
   (q) $\frac{1}{2}$ (r) $\frac{1}{2}$ (s) $\frac{1}{\sqrt{2}}$ (t) $-\frac{1}{\sqrt{3}}$

3. (a) 19.4°, 160.6°   (b) 132.2°, 312.2°
   (c) 138.5°, 221.5°   (d) 21.0°, 201.0°
   (e) 20.4°, 299.6°   (f) 76.4°, 256.4°
   (g) 14.2°, 75.8°, 194.2°, 255.8°
   (h) 19.6°, 100.4°, 139.6°, 220.4°, 259.6°, 340.4°
   (i) 29.7°, 80.3°, 260.3°, 209.7°
   (j) 50.7°, 269.3°   (k) 31.4°, 127.0°
   (l) 23.8°, 83.8°, 143.8°, 203.5°, 263.5°, 323.5°
   (m) 30°, 150°   (n) 90°, 270°, 48.6°, 131.4°
   (o) 14.5°, 165.5°, 30°, 150°
   (p) 135°, 315°

4. 0.99ᶜ, 2.15ᶜ

5. (a) 135°   (b) 105°, 165°

6. (a) 2.50ᶜ, 3.79ᶜ   (b) $\frac{\pi}{2}, \frac{7\pi}{6}$

Exercise 7E

1. (a) $22.8°$, $157.2°$, $202.8°$, $337.2°$
   (b) $67.2°$, $292.8°$, $112.8°$, $247.2°$
   (c) $36.3°$, $323.7°$, $143.8°$, $216.3°$
2. (a) $91°$ (b) $215°$
3. (a) $66°$, $204°$ (b) $53°$, $307°$
   (c) $72.5°$, $287.5°$
4. missing values 2.87, 2.83; $38 < x < 68$
5. -0.62, -1, -0.62, 0.25, 1; $38.2°$, $141.8°$
8. $P(1.6, 1)$, $Q(-1.6, -1)$

9. (b) $\dfrac{\pi}{3}, \dfrac{2\pi}{3}, \dfrac{4\pi}{3}, \dfrac{5\pi}{3}, \dfrac{7\pi}{3}, \dfrac{8\pi}{3}$

Exercise 8A

1. y-coordinate: (a) 8 (b) 3.375 (c) 1.331
   (d) $1.030\,301$ (e) $1.000\,300$
   gradient PQ: (a) 7 (b) 4.75 (c) 3.31
   (d) 3.0301 (e) 3.0000
   Gradient of tangent to curve at P is 3
2. (a) -2 (b) $\frac{1}{2}$ (c) $-\frac{1}{3}$
3. (a) $4x^3$ (b) $-3x^{-4}$ (c) 0 (d) 3
   (e) $12x^2$ (f) $-10x^{-6}$ (g) $-\dfrac{1}{2x^2}$
   (h) $-\dfrac{3}{x^3}$ (i) $2x + 2x^{-3}$
   (j) $-3x^{-2} + 4x^{-3}$ (k) $2x + 1$
   (l) $3x^2 - 3$ (m) 1 (n) $16x^3 - 24x$
   (o) $x + \dfrac{1}{2x^2}$
4. (a) -6 (b) -12 (c) -3 (d) $-\frac{1}{16}$
   (e) $3\frac{1}{4}$ (f) 8 (g) $-1\frac{1}{4}$
5. $2, 0, -2$ 6. $-12, -3, -12, -3$
7. $y = 1$, $\dfrac{dy}{dx} = -5$
8. $A(100, 0)$, -100. $H(50, 2500)$. H is the highest point reached by the arrow in its flight.

Exercise 8B

1. (a) $4e^{4x}$ (b) $7e^{7x}$ (c) $-6e^{-2x}$
   (d) $5e^{-5x}$ (e) $-2e^{-\frac{x}{3}}$ (f) $-2e^{-2x}$
   (g) $-6e^{-\frac{3x}{2}} + 2$ (h) $3e^{3x} + 3e^{-3x}$
   (i) $2e^{2x} - 2e^x$ (j) $2e^{2x}$
   (k) $2x - 3e^{3x}$ (l) $-3e^{-3x} + 5e^{-5x}$
   (m) $-e^{-x} + 3e^{-3x}$
2. (a) 403, 1210 (b) 0.368, 0.184
   (c) 3.09, 2.35 (d) 2.95, 3.30
   (e) 0, 8
3. $(0, 1)$, 0.69, -1.10

Exercise 8C

1. (a) $\dfrac{2}{x}$ (b) $\dfrac{1}{x}$ (c) $\dfrac{4}{x} + \frac{1}{2}x^{-\frac{1}{2}}$
   (d) $\dfrac{1}{2x} + \dfrac{x}{2}$ (e) $\dfrac{1}{3x} + e^{\frac{1}{2}x}$ (f) $-\dfrac{5}{x}$
   (g) $\dfrac{1}{x} + 6x$ (h) $-2e^{-2x} - \dfrac{5}{x}$
   (i) $-\dfrac{2}{x^2} - \dfrac{3}{x}$ (j) $\dfrac{4}{x} + 3x^{-2}$
2. (a) 1.69, 2.39 (b) 0.5, 0.25
3. (a) 2 (b) $4\frac{1}{6}$ (c) -5 (d) -3.28
   (e) 0.35
4. (a) 1 (b) $\frac{1}{3}$ (c) 2
5. (a) $f^{-1} : x \mapsto \ln(x - 1)$, $x \in \mathbb{R}$, $x > 1$
   (c) $\dfrac{df}{dx} : x \mapsto e^x$, $\dfrac{df^{-1}}{dx} : x \mapsto \dfrac{1}{x - 1}$, $x > 1$

Exercise 8D

1. $-3x^2$; f is decreasing because $f'(x) < 0$
2. (a) (i) $x > 2$ (ii) $x < 2$
   (b) (i) $x > 0$ (ii) $x < 0$
   (c) (i) $x < 0$ (ii) $x > 0$
   (d) (i) $x < 1$ (ii) $x > 1$
3. $\frac{3}{4}$; $-\frac{3}{2}$
4. greatest value is 4 at $x = 2$; $y \leqslant 4$
5. (a) $(0, 0)$ min, $\left(-\frac{2}{3}, \frac{4}{27}\right)$ max

(b) (1, 0) min, (−1, 4) max
(c) (1, 2) min, (−1, −2) max
(d) (2.38, 17.0) min
(e) $(\frac{2}{3}, -\frac{25}{27})$ min, $(-\frac{1}{2}, \frac{9}{4})$ max
(f) (2, 2) min, (0, 6) max

6 (1, −1) min, (−2, 26) max

7 Cuts x-axis where $x = -\sqrt{3}, 0$ and $\sqrt{3}$. Max. at (1, 2); min. at (−1, −2)

8 $\frac{1}{3}$ and −2

9 (a) (1, 4) and $(1\frac{2}{3}, 3\frac{5}{27})$ (b) $\frac{2}{3} < x < 2$

10 (a) (1.39, −1.55)   (b) $x > 1.39$

11 $x > 2$ **12** 0 min, −4 max

13 500 m, 10

14 a minimum value of 12 at $x = 2$

15 2500   **16** 20   **17** 125 000

18 3.30, 65.4   **19** 4   **20** 16π cm³, 2 cm

21 18 000 cm³   **22** 13.86 cm $(4\sqrt{12}$ cm)

23 0.1 cm s⁻¹; 0.3 cm s⁻¹; $A = \dfrac{\pi t^4}{100}$; 0.016 cm² s⁻¹

0.424 cm² s⁻¹

24 $V = \dfrac{\pi t^3}{6}$; $A = \pi t^2$; $\dfrac{9\pi}{2}$ cm³ s⁻¹; 6π cm² s⁻¹

Exercise 9A

1 (a) $\dfrac{x^2}{2} + C$ (b) $\dfrac{x^5}{5} + C$ (c) $\dfrac{3x^4}{2} + C$

(d) $-x^{-1} + C$ (e) $-\frac{2}{3}x^{-3} + C$

(f) $3x^3 + C$ (g) $2x^2 + 5x + C$

(h) $\dfrac{x^3}{3} - \dfrac{x^2}{2} + C$ (i) $x - \dfrac{x^2}{2} + C$

(j) $\dfrac{x^3}{3} + x^2 + x + C$

(k) $4x - 2x^2 + \dfrac{x^3}{3} + C$

(l) $\dfrac{x^3}{3} - 2x - \dfrac{1}{x} + C$

(m) $-x^{-1} - \dfrac{x^3}{3} + C$

(n) $x - 2x^{-1} - \frac{1}{3}x^{-3} + C$

(o) $\dfrac{x^6}{2} + x^{-2} + C$ (p) $-5x^{-3} + C$

(q) $3x^3 - 12x - 4x^{-1} + C$

(r) $2x^3 - \frac{13}{2}x^2 + 6x + C$

(s) $5x^2 - 9x + C$ (t) $\dfrac{x^3}{3} - \dfrac{8}{5}x^{\frac{5}{2}} + 2x^2 + C$

2 (a) $\frac{2}{3}x^{\frac{3}{2}} + C$ (b) $\frac{5}{8}x^{\frac{8}{5}} + C$ (c) $2x^{\frac{1}{3}} + C$
(d) $4x^{\frac{1}{4}} + C$ (e) $2x^{\frac{5}{2}} + C$ (f) $\frac{2}{5}x^{\frac{5}{2}} + C$
(g) $-2x^{-\frac{1}{2}} + C$ (h) $3x^{\frac{4}{3}} + C$ (i) $\frac{4}{3}x^{\frac{9}{4}} + C$

(j) $2x^{-\frac{1}{4}} + C$ (k) $\dfrac{2\sqrt{3}}{3}x^{\frac{3}{2}} + C$

(l) $x^{\frac{1}{2}} + C$

3 (a) $\frac{1}{7}e^{7x} + C$ (b) $-\frac{1}{3}e^{-3x} + C$
(c) $e^{2x} + C$ (d) $e^{\frac{x}{2}} + C$
(e) $\frac{1}{2}e^{2x} - 3x + C$ (f) $-2e^{-3x} - e^{3x} + C$
(g) $-\frac{1}{4}e^{4x} - e^{2x} + x + C$ (h) $e^x + e^{-x} + C$
(i) $2x^3 - 2e^{-2x} + C$
(j) $-\frac{3}{2}e^{-6x} - \frac{3}{2}e^{-4x} - \frac{1}{2}e^{-2x} + C$

4 (a) $\ln x + 2x + C$ (b) $x^2 - \ln x + C$
(c) $2\ln x + x^{-1} + C$ (d) $3\ln x + C$
(e) $\frac{1}{3}\ln x + C$ (f) $\frac{5}{8}\ln x + C$
(g) $-9x^{-1} - 6\ln x + x + C$

(h) $2x^2 + \ln x + C$ (i) $\dfrac{3x^2}{4} - 2\ln x + C$

(j) $6x + 5\ln x + 6x^{-1} + C$

Exercise 9B

1 (a) $y = x^2 - x + 1$ (b) $y = \dfrac{x^4}{4} + \dfrac{x^2}{2} + 1$

(c) $y = e^x + x^2$ (d) $y = \frac{3}{2} - \frac{1}{2}e^{-2x}$
(e) $y = 3x^3 - 12x^2 + 16x + 1$
(f) $y = 2e^{\frac{x}{2}} - 1$

2 (a) 8 (b) $\frac{1}{4}$ (c) $\frac{14}{3}$ (d) $\frac{14}{3}$
(e) $e - 1$ (f) $3[\ln 5 - \ln 2]$

4 $-\frac{1}{6}$ **5** $31\frac{1}{3}$ **6** −18

7 $\frac{1}{2}e^4 - 4e^2 + \frac{23}{2}$ **8** $\frac{2}{7}, 0, \frac{2}{9}, 0$

9 $y = \frac{1}{4}x^4 - \frac{1}{2}x^2 + \frac{17}{4}$; $\frac{65}{8}$ **10** $5\frac{1}{2}$

11 (i) $130\frac{4}{5}$ (ii) $-\frac{62}{81}$ **12** 2, 8

Exercise 9C

1 (a) 42　(b) $36\frac{2}{7}$　(c) $1\frac{1}{3}$　(d) $e^3 - 1$
(e) $\frac{3}{4}\ln 4$　(f) 2　　(g) 204 (3 s.f.)

2 $(5, 0)$; $20\frac{5}{6}$　**3** 24　**4** 4　**5** $4\frac{1}{2}$

6 21; sign positive, indicating region is above the
x-axis; -21

7 area $4\frac{1}{2}$　**8** $-\frac{1}{4}, \frac{1}{4}, 0$

9 $(1, 0)$, $(4, 0)$: regions are congruent

10 $\frac{1}{3}$　**11** $\frac{1}{6}$　**12** 1.69　**13** $78\frac{2}{3}$

14 $8\frac{1}{3}$　**15** $3\frac{21}{32}$　**16** 16　**17** 8.93

18 154.48　**19** $\dfrac{e^2}{2} + \dfrac{5}{2e}$

20 (i)　$(0, -4)$ min $(-2, 0)$ max　(ii) $2\frac{3}{4}$

Exercise 10A

1 292.5, 293.5, 0.5　**2** 155, 165, 5

3 1.1235, 1.1245, 0.0005

4 25 500, 26 500, 500　**5** 115, 125, 5

6 0.01255, 0.01265, 0.00005

7 0.425, 0.435, 0.005　**8** 8961.5, 8962.5, 0.5

9 1 324 167.5, 1 324 168.5, 0.5

10 1 250 000, 1 350 000, 50 000

Exercise 10B

1 0.0019　**2** 0.5　**3** 0.0015　**4** 0.0263

5 0.0357　**6** 0.0036　**7** 0.5　**8** 0.0001

9 0.0032　**10** 0.25　**11** 0.025

12 0.0002

Exercise 10C

1 (a) 14.6, 14.4　(b) 12, 11.8
(c) 17.8875, 16.4375　(d) 10.6, 9.74
(e) 0.35, 0.33　(f) 212.75, 183.75
(g) 2.47, 1.9　(h) 41.8125, 40.6525
(i) 77.2, 65.1　(j) 6.4, 6.2

(k) 11.4, 10.1　(l) 26.66, 26.65

2 (a) $0.0505, 3.11 \times 10^{-3}$
(b) $0.505, 4.15 \times 10^{-4}$
(c) $0.0055, 6.07 \times 10^{-4}$　(d) $1, 7.35 \times 10^{-3}$
(e) $0.555, 5.19 \times 10^{-4}$
(f) $0.6, 3.80 \times 10^{-3}$　(g) 1.13, 0.0173
(h) 0.35, 0.174　(i) 1.08, 0.169
(j) 0.682, 0.0445　(k) 9.05, 0.0449
(l) 23.45, 0.0120　(m) 0.154, 0.0112
(n) 0.154, 0.0112

3 (a) 25.4, 25.0　　(b) 33.5625, 32.3025

4 1.41, 2.2; 0.0403, 0.0603; 0.011, 0.0191

5 (a) $10.78\,\text{cm}^2$　(b) $10.31\,\text{cm}^2$　(c) $7\,\text{cm}$
(d) $17\,\text{cm}$　(c) $15.8\,\text{cm}$

Exercise 10D

6 $a = 22$　**12** 1　**15** (b) 1　**16** 2.2

Review exercise 3

1 $2, -3$　**2** -1　**3** $-145°, 35°, 215°$

4 (a) $y = \frac{12}{5}\ln x + 2$　(b) 7.53

5 $-4, 6$　**6** $p = -3, q = 3$　**7** $-4\frac{1}{2}$

8 $85\frac{1}{3}$　**9** -4.48　**10** $\frac{9}{16}$

11 (a) $44.4°, 135.6°$　(b) $\pm45.6°$
(c) $35°, -145°$

12 (a) 3.67, 5.76　(b) 0.64, 5.64
(c) 1.25, 4.39

13 (a) $45.5\,\text{cm}$　(b) $106\,\text{cm}^2$

14 (a) $8.62\,\text{cm}^2$　(b) $7.90\,\text{cm}$

15 (a) 2.24　(b) 1.4

16 (a) $(-1, 1), (3, 5)$　(b) $5\frac{1}{3}$　(c) $y \geqslant \frac{1}{2}$
(d) $x \geqslant 0$

17 (a) 2, 3　(b) 5

18 $(2, 20)$; $(4, 16)$. The line $y = 18$ crosses curve 3
times; 36

19 $57\frac{1}{6}$, $y \leqslant 12\frac{1}{4}$

20 $\left(135°, \frac{\sqrt{2}}{2}\right)$, $\left(315°, -\frac{\sqrt{2}}{2}\right)$

21 (a) 29.6 cm (b) 52.5675 cm²

22 8.68 m s⁻¹ to 9.10 m s⁻¹ (3 s.f.)

23 6.75 and 13.75, 1.8 and 3.67

24 (b) $64 - 64x + 12x^2$ (c) $1\frac{1}{3}$ (e) $37\frac{25}{27}$

25 (a) 10°, 50° (b) 0, 45°

26 $\left(-\frac{\pi}{6}, 1\right)$ **27** $2y = 2x^3 - e^{2x} - 1$

28 $A = 2\pi x^2 + 32\pi x^{-1}$; 2, 24π cm²

29 $32\sqrt{2}$, $\frac{512}{5}\sqrt{2}$

30 $A = \pi(5t^2 + 4)^2$; 30 cm s⁻¹, 2940π cm² s⁻¹

31 $4\frac{1}{2}$ **32** $\frac{1}{216}$ **33** $2\frac{2}{3}$

34 (i) 4 (ii) 28.2 (iii) $2e - \frac{2}{\sqrt{e}}$

35 (a) 4 (c) $4.5 - 6\ln 2$ (d) $-2, 1$

36 170.5π

37 (a) 3.146 cm, $51.5° \leqslant C \leqslant 54.9°$

38 (i) 0.05 (ii) 0.008 77

39 928.36 cm³ **40** 24.34 cm³

41 (a) 15.6 cm² (b) 25.4 cm²

41 (b) 10 (c) 5 cm

43 (a) 20π cm (b) $100(\pi - 1)$ cm²
  (c) 45.5 cm²

44 $6\frac{3}{4}$

45 (a) $\frac{\pi}{18}, \frac{5\pi}{18}, \frac{13\pi}{18}, \frac{17\pi}{18}$ (b) $\frac{3\pi}{4}$

46 (b) $\frac{a^3}{6}$ (c) $\left(\frac{a}{3}, \frac{2a^2}{9}\right)$, $\left(\frac{2a}{3}, \frac{2a^2}{9}\right)$

(d) $\frac{a^3}{162}$

47 (b) $V = 100x - \frac{4}{3}x^3$ (c) $333\frac{1}{3}$ cm³

48 (a) (6, 12) (b) $13\frac{1}{3}$

49 $k = -4$; maximum $y = 6$

50 (b) (4, 16) (c) 133 (3 s.f.)

Examination style paper P1

1 $\frac{q^2 + pq}{p^2 - p^4 q}$ **2** 0, 3

3 (a) $y - 2 = 3(x - 3)$
  (b) $y = 13 - 2x$; (4, 5)

4 $x < -1$ and $x > 2$

5 (a) 5.85, -0.85 (b) $x > 6$ or $x < -1$

6 (a) $A(1, 0)$, $B(5, 0)$, $C(6, 5)$ (b) $10\frac{1}{6}$

7 (a) 320, $\frac{1}{2}$ (c) converges to the value 640

8 (b) 30 000, 100

9 (a) $f(x) \geqslant -1$, $y = x^2 - 4x + 3$
  (c) (i) $(-1, 0)$, $(1, 0)$; $(0, -1)$ min
    (ii) $(0, 3)$, $(\frac{1}{2}, 0)$, $(\frac{3}{2}, 0)$; $(1, -1)$ min

10 (b) $A(0, 1)$, $B\left(45, \frac{\sqrt{2}}{2}\right)$, $C\left(225, -\frac{\sqrt{2}}{2}\right)$

  (d) 3; 30°, 150°, 270° (e) $y = \cos x$

List of symbols and notation

The following symbols and notation are used in the London modular mathematics examinations:

$\{ \quad \}$	the set of
$n(A)$	the number of elements in the set A
$\{x : \quad \}$	the set of all x such that
$\in$	is an element of
$\notin$	is not an element of
$\emptyset$	the empty (null) set
$\mathscr{E}$	the universal set
$\cup$	union
$\cap$	intersection
$\subset$	is a subset of
A'	the complement of the set A
PQ	operation Q followed by operation P
$f : A \rightarrow B$	f is a function under which each element of set A has an image in set B
$f : x \mapsto y$	f is a function under which x is mapped to y
$f(x)$	the image of x under the function f
f^{-1}	the inverse relation of the function f
fg	the function f of the function g

open interval on the number line

closed interval on the number line

$\mathbb{N}$	the set of positive integers and zero, $\{0, 1, 2, 3, \ldots\}$
$\mathbb{Z}$	the set of integers, $\{0, \pm1, \pm2, \pm3, \ldots\}$
$\mathbb{Z}^+$	the set of positive integers, $\{1, 2, 3, \ldots\}$
$\mathbb{Q}$	the set of rational numbers
$\mathbb{Q}^+$	the set of positive rational numbers, $\{x : x \in \mathbb{Q}, x > 0\}$
$\mathbb{R}$	the set of real numbers
$\mathbb{R}^+$	the set of positive real numbers, $\{x : x \in \mathbb{R}, x > 0\}$
$\mathbb{R}_0^+$	the set of positive real numbers and zero, $\{x : x \in \mathbb{R}, x \geqslant 0\}$
$\mathbb{C}$	the set of complex numbers
$\sqrt{}$	the positive square root
$[a, b]$	the interval $\{x : a \leqslant x \leqslant b\}$
$(a, b]$	the interval $\{x : a < x \leqslant b\}$
(a, b)	the interval $\{x : a < x < b\}$

$\lvert x \rvert$	the modulus of $x = \begin{cases} x \text{ for } x \geqslant 0 \\ -x \text{ for } x < 0 \end{cases}, x \in \mathbb{R}$
$\approx$	is approximately equal to
A^{-1}	the inverse of the non-singular matrix A
A^T	the transpose of the matrix A
$\det A$	the determinant of the square matrix A
$\displaystyle\sum_{r=1}^{n} f(r)$	$f(1) + f(2) + \ldots + f(n)$
$\displaystyle\prod_{r=1}^{n} f(r)$	$f(1)f(2) \ldots f(n)$
$\begin{pmatrix} n \\ r \end{pmatrix}$	the binomial coefficient $\dfrac{n!}{r!(n-r)!}$ for $n \in \mathbb{Z}^+$ $\dfrac{n(n-1)\ldots(n-r+1)}{r!}$ for $n \in \mathbb{Q}$
$\exp x$	e^x
$\ln x$	the natural logarithm of x, $\log_e x$
$\lg x$	the common logarithm of x, $\log_{10} x$
arcsin	the inverse function of sin with range $[-\pi/2, \pi/2]$
arccos	the inverse function of cos with range $[0, \pi]$
arctan	the inverse function of tan with range $(-\pi/2, \pi/2)$
arsinh	the inverse function of sinh with range $\mathbb{R}$
arcosh	the inverse function of cosh with range $\mathbb{R}_0^+$
artanh	the inverse function of tanh with range $\mathbb{R}$
$f'(x), f''(x), f'''(x)$	the first, second and third derivatives of $f(x)$ with respect to x
$f^{(r)}(x)$	the rth derivative of $f(x)$ with respect to x
$\dot{x}, \ddot{x}, \ldots$	the first, second, . . . derivatives of x with respect to t
z	a complex number, $z = x + iy = r(\cos\theta + i\sin\theta) = re^{i\theta}$
$\text{Re } z$	the real part of z, $\text{Re } z = x = r\cos\theta$
$\text{Im } z$	the imaginary part of z, $\text{Im } z = y = r\sin\theta$
z^*	the conjugate of z, $z^* = x - iy = r(\cos\theta - i\sin\theta) = re^{-i\theta}$
$\lvert z \rvert$	the modulus of z, $\lvert z \rvert = \sqrt{(x^2 + y^2)} = r$
$\arg z$	the principal value of the argument of z, $\arg z = \theta$, where $\left.\begin{matrix} \sin\theta = y/r \\ \cos\theta = x/r \end{matrix}\right\} - \pi < \theta \leqslant \pi$
$\mathbf{a}$	the vector $\mathbf{a}$
$\overrightarrow{AB}$	the vector represented in magnitude and direction by the directed line segment AB
$\hat{\mathbf{a}}$	a unit vector in the direction of $\mathbf{a}$
$\mathbf{i,j,k}$	unit vectors in the directions of the cartesian coordinate axes
$\lvert \mathbf{a} \rvert$	the magnitude of $\mathbf{a}$
$\lvert \overrightarrow{AB} \rvert$	the magnitude of $\overrightarrow{AB}$
$\mathbf{a.b}$	the scalar product of $\mathbf{a}$ and $\mathbf{b}$
$\mathbf{a} \times \mathbf{b}$	the vector product of $\mathbf{a}$ and $\mathbf{b}$

A'	the complement of the event A
$P(A)$	probability of the event A
$P(A|B)$	probability of the event A conditional on the event B
$E(X)$	the mean (expectation, expected value) of the random variable X
$X, Y, R,$ etc.	random variables
$x, y, r,$ etc.	values of the random variables $X, Y, R,$ etc.
$x_1, x_2 \ldots$	observations
$f_1, f_2, \ldots$	frequencies with which the observations $x_1, x_2, \ldots$ occur
$p(x)$	probability function $P(X = x)$ of the discrete random variable X
$p_1, p_2, \ldots$	probabilities of the values $x_1, x_2, \ldots$ of the discrete random variable X
$f(x), g(x), \ldots$	the value of the probability density function of a continuous random variable X
$F(x), G(x), \ldots$	the value of the (cumulative) distribution function $P(X \leqslant x)$ of a continuous random variable X
$\mathrm{Var}(X)$	variance of the random variable X
$B(n, p)$	binomial distribution with parameters n and p
$N(\mu, \sigma^2)$	normal distribution with mean μ and variance σ^2
μ	population mean
σ^2	population variance
σ	population standard deviation
$\bar{x}$	sample mean
s^2	unbiased estimate of population variance from a sample,

$$s^2 = \frac{1}{n-1} \sum (x - \bar{x})^2$$

ϕ	probability density function of the standardised normal variable with distribution $N(0, 1)$
Φ	corresponding cumulative distribution function
α, β	regression coefficients
ρ	product-moment correlation coefficient for a population
r	product-moment correlation coefficient for a sample
$\sim p$	not p
$p \Rightarrow q$	p implies q (if p then q)
$p \Leftrightarrow q$	p implies and is implied by q (p is equivalent to q)

Index